OCEAN FLOOR MINING

OCEAN FLOOR MINING

John S. Pearson

NOYES DATA CORPORATION
Park Ridge, New Jersey   London, England
1975

Copyright © 1975 by Noyes Data Corporation
    No part of this book may be reproduced in any form
    without permission in writing from the Publisher.
Library of Congress Catalog Card Number: 75-801
ISBN: 0-8155-0569-8
Printed in the United States

Published in the United States of America by
Noyes Data Corporation
Noyes Building, Park Ridge, New Jersey 07656

FOREWORD

This Ocean Technology Review contains carefully edited and collated data culled from many scattered and difficult-to-obtain sources. The geographical and environmental information is based on studies carried on by research teams under the auspices of various governmental agencies. Proceedings of undersea mining symposiums were the source of the legal and economic considerations involved in undersea mining operations. The technological studies have been obtained from a variety of sources which include governmental studies and private proprietary industrial research by companies interested in the acquisition of the mineral wealth which lies at the bottom of the world's oceans.

Advanced composition and production methods developed by Noyes Data are employed to bring new durably bound books to the reader in a minimum of time. Special techniques are used to close the gap between "manuscript" and "completed book." Industrial technology is progressing so rapidly that time-honored, conventional typesetting, binding and shipping methods are no longer suitable. Delays in the conventional book publishing cycle have been bypassed to provide the user with an effective and convenient means of reviewing up-to-date information in depth.

The Table of Contents is organized in such a way as to serve as a subject index and provides easy access to the information contained in this book.

CONTENTS AND SUBJECT INDEX

INTRODUCTION

On the ocean floor are vast quantities of mineral reserves; their exploitation being limited primarily by the technology of their recovery or delivery to the surface of the ocean. Among the primary mineral resources presently known are rich deposits of zinc, copper, silver, lead, manganese and phosphate.

There are two basic varieties of mineral deposits on the ocean floor that are potential ores: first, ore deposits of hydrothermal origin associated with the worldwide ocean ridge system; second, concentrations of minerals probably derived from seawater and spread as a thin layer or as nodules over large areas of the ocean floor. The latter concentrations are the primary potential sources of manganese and phosphate while the former deposits associated with the world ridge system represent potential concentrations of copper, zinc, lead, gold, and silver.

Ocean mining offers many advantages which are not possible with traditional land mining. In the ocean there are materials that are available without removing any overburden, without the use of explosives, and without the expense of drilling operations for sampling. With cameras and inexpensive coring operations the complete deposit can be explored prior to mining.

As a whole new concept, ocean mining can be designated for automation in the beginning which should result in new equipment designs not bound by tradition. The same equipment could be used to mine various deposits and could be easily moved from one area to another. Sea transportation can be used to carry the mined ore to more of the world's markets with no other form of transportation involved.

The midoceanic ridge system is a long ridge extending for some 40,000 miles through the main oceans of the world. Associated with its crest is a narrow zone where volcanic activity is concentrated. This is an area of high frequency of shallow earthquakes. Iceland is a part of this ridge which is emergent, and the volcanic activity and hot spring activity of Iceland is probably typical of very long reaches of this ridge.

Experts have hypothesized that the crestal region of this ridge probably contains such vast quantities of heavy metal deposits that it may revolutionize the whole heavy metal industry. The metals, apparently, are derived from hot waters emerging from deep within the earth carrying large amounts of mineral materials, and these minerals are precipitated in the sediment as the waters percolate through them. Two deposits of this type have been found in the deep ocean, and a third found off Southern California is associated with the same geological feature.

The most thoroughly investigated of these deposits is an area of heavy metals in the Red Sea. In 1964 two subsurface pools of hot saline brine were discovered on the floor of the Red Sea, and a third small pool was found in 1966. The pools occur in adjacent local depressions along the medium valley of the Red Sea. The brines in these pools have heavy metal concentrations that are much above normal ocean water, and their associated bottom sediments contain the highest contents of zinc, lead and copper yet found in recent marine deposits.

At current smelter prices for zinc, copper, lead, silver and gold, the metals of these deposits have been conservatively estimated as worth about 2.3 billion dollars. Accordingly, impetus for providing equipment capable of economically recovering these ores is clearly present.

Large areas of the Pacific Ocean contain vast fields of manganese nodules, and the only obstacles to their use as an ore is the cost of recovery and transportation to some suitable refinery. A recent discovery of a manganese crust on the Blake Plateau is of particular interest because of its proximity to the United States and suitable areas for refining the material and marketing it.

Dredge samples and photographs from the Blake Plateau off the southeast coast of the United States indicate that a layer of manganese oxide forms a pavement that may be continuous over an area of about 5,000 square kilometers. The manganese pavement grades into round manganese nodules to the south and east and into phosphate nodules to the west. The Gulf Stream probably maintains a very unusual environment that prohibits deposition of other sediment on top of this deposit and permits the accretion of the manganese pavement.

The Pacific Ocean floor is estimated to contain 1.5 trillion tons of manganese nodules, which also are rich in copper, nickel, cobalt and other metals. In addition, several important materials can be recovered from the seas ranging from salt to metal-bearing sands.

But before these riches can be tapped, several thorny questions must be answered: how far out does a nation's jurisdiction over its coastal resources extend? Should there be a uniform worldwide territorial limit? Who owns deep-ocean minerals lying far from any coast? What are the rights of passage of one country's ships through territorial waters of another? What are the limits of scientific research on the high seas? And how can the vast wealth of the oceans be brought up at minimum impact on the environment?

There is an expanding market for the metals found in these nodules. For instance, because of growing demand the U.S. in 1970 imported about 18% of the copper needed, 74% of the nickel, 95% of the manganese and 98% of the cobalt. And this increasing demand has spurred interest in improving technology to the point where engineering advances in the recovery and processing of the nodules have made exploitation of the minerals economically feasible.

Once decisions to press forward with commercial nodule development are made, companies will be faced with the task of collecting the nodules in economical quantities. Several recovery methods have been devised. Also getting attention is the processing of the nodules, extracting the relatively pure metals from them.

In anticipation of the deep-sea mining of manganese nodules, the U.S. National Oceanic and Atmospheric Administration has begun efforts to determine the environmental consequences. About 850 miles southeast of Honolulu, a team of scientists has been gathering data on the sediment overlaying the nodules and the water column above it. The information gained and the experiments conducted will provide baselines for later monitoring of prototype mining operations.

Although manganese nodules now occupy center stage on the undersea minerals scene, there are other promising prospects. Common salt, magnesium and bromine have long been extracted from seawater itself and doubtless will continue to be. Other minerals, such as sand and gravel, tin-bearing sands, calcium carbonate and magnetite (iron) sands possibly will be mined from the sea, although only tin is likely to prove commercially exploitable.

Another potentially valuable offshore mineral is barite (barium sulfate), used as an additive to drilling muds. Promising deposits of the material have been found off Alaska, but no large-scale development is in the works. Potential gold mines of mineral wealth are the hot brines of the Red Sea. These high-temperature mineral solutions welling up from the sea bottom have been found to contain substantial amounts of copper, silver, zinc, antimony and other valuable metals.

This book represents a sincere attempt to gather together and collate the available information concerning the state-of-the-art of deep-sea mining. We hope that it may prove valuable to those interested in tapping these ocean resources.

GEOGRAPHY OF
FERROMANGANESE NODULE DEPOSITS

The information in this section concerning the worldwide distribution and metal content of ferromanganese nodule deposits in the world oceans was obtained from the following sources:

A report presented by D.R. Horn, B.M. Horn and M.N. Delach at the conference on Ferromanganese Deposits on the Ocean Floor held at Arden House, Harriman, New York and Lamont-Doherty Geological Observatory, Columbia University, Palisades, New York, January 20-22, 1972 from work done under IDOE Grant GX 30675, ONR Contract No. 0014-67-A0108-004 and NSF Grant GA 29460 and issued as *PB 225 985;*

an article by D.R. Horn, M. Ewing, B.M. Horn and M.N. Delach published in *Ocean Industry*, January 1972, pp 26-29 based on work done under IDOE Grant GX 30675, ONR Contract No. 0014-67-A-0108-0004 and NSF Grants GA 27281 and GA 29460;

and Technical Report No. 3 by D.R. Horn, M.N. Delach and B.M. Horn under IDOE/NSF Grant GX 33616 issued as *PB 223 083* (unpublished manuscript).

WORLDWIDE DISTRIBUTION AND METAL CONTENT IN WORLD OCEANS

The widespread occurrence of ferromanganese nodules and crusts on the ocean floor was established 100 years ago during the deep-sea expedition of H.M.S. *Challenger* (Murray and Renard, 1891). Since then several hundred reports of similar deposits have been made and these data have been compiled and reproduced on maps.

There are two dominant types of ferromanganese deposits each reflecting conditions at the site of deposition. Encrusting material develops on exposed submarine elevations where current activity prevents normal sediment accumulation.

4

Currents provide a continuous supply of metals which accrete to exposed surfaces and form ferromanganese crusts. The second type is nodular and forms at great depth where sediment accumulation is negligible. In these deep abyssal areas, ferromanganese precipitates about nuclei (volcanic, biologic, glacial, nodule fragment, etc.) and in time the addition of concentric layers result in the formation of nodules.

A compilation of published and unpublished chemical analyses of samples of ferromanganese deposits from the ocean floor reveals that only in relatively few areas of the world are the copper and nickel contents sufficiently high for the nodules to represent a potential source of metals.

The North and South Atlantic and Indian Oceans are characterized by deposits whose metal content (Cu, Ni, Mn and Co) are well below the minimum values necessary for economic exploitation.

The situation is more favorable in the Pacific Ocean and most encouraging in the North Pacific. In the South Pacific, nodules containing over 1% Ni and lesser amounts of copper occur in the Peru Basin, in deep waters east of the Marquesas Islands and Tuamotu Plateau, and within the Southwest Pacific Basin. The relatively low Cu and Ni values obtained in these areas may eliminate them as prospective mining sites.

Only in the North Pacific do the analyses consistently show values greater than 1% Cu and 1% Ni, and these nodules, therefore, are the immediate target of the ocean mining industry. The deposits of interest lie north of the equator in a broad band between 6° 30' N and 20° N and stretching from 110° W to 180° W. Maximum copper and nickel values occur along a line 8° N to 10° 30' N and 131° 30' W to 145° W.

North Pacific Ocean

The North Pacific offers a favorable picture as a potential area for mining nodules. Major markets are at hand in highly industrialized countries such as Japan and the United States. Data reveal a high concentration of manganese nodules within an east-west band which stretches across the Pacific from deep waters off Central America to the seaward wall of the Mariana Trench. Northern and southern limits of the zone are 20° N and 6° 30' N, respectively.

Fewer nodules occur in the western half of the area because of abundant chains of major seamounts. The latter are local sources of sediment whose relatively rapid rate of deposition precludes development of nodules. They are abundant in deepest parts of intermountain basins. Data confirm that nodules are abundant immediately north of the thick pile of biogenic sediments of the Equatorial Pacific.

The distribution of nodules is clearly related to the southern edge of the vast region of red clay deposition which dominates much of the North Pacific Basin. This province is characterized by rolling abyssal hills covered by very fine-grained sediment. Some properties of the deposits at the water-sediment interface include:

average particle sizes range from 0.86 to 1.73 microns (average 1.08 microns);

porosities from 70 to 90% (average 80%);
moisture contents from 95.21 to 334.64% of dry weight (average 176.86%);
wet densities from 1.18 to 1.48 g/cc (average 1.33 g/cc); and
void ratios from 2.41 to 9.72 (average 4.63).

The nodules are found in depths between 3,206 and 5,997 meters (average 4,853 meters). Rates of sediment accumulation are extremely low and range from less than 1 mm/1,000 years to 3 mm/1,000 years (Opdyke and Foster, 1970). Sand or granule-sized volcanic material either in the form of basaltic glass or its alteration products serve as the principal nuclei of manganese precipitation.

North of 20° N latitude there are fewer nodules. This may be due in part to lower density of available core data. However, it is more likely a function of slightly higher rates of deposition. Areas of red clay deposition in very deep water which are beyond the influence of biogenic and terrigenous sources of sediment appear to offer the conditions necessary for nodule development. Within the North Pacific Ocean the highest concentration of nodules lies between 20° N and 6° 30' N.

The North Pacific Ocean is the largest sedimentary basin in the world. It receives very little land-derived sediment. Continental debris is trapped at the periphery of the basin. There is no mid-ocean ridge system to serve as a major source of sediment. Therefore, as above stated most of the North Pacific Basin is characterized by very slow sedimentation. It is the site of widespread and intense development of nodular deposits of ferromanganese.

Their distribution is a function of low rate of deposition of red clay (less than 1 to 3 mm/1,000 years) and of siliceous ooze (3.5 mm/1,000 years). Based on information in hand, the North Pacific has the highest density of nodular deposits in the world ocean with most occurring within an east-west band between the abovementioned limits of 6° 30' N and 20° N.

The North Pacific is not like the Atlantic and Indian Oceans in that continental debris is trapped in secondary basins at the periphery of the ocean and there is no large mid-ocean ridge system over which carbonate sediment can accumulate and be resedimented in adjacent deep water. The Emperor Seamount Chain and the Hawaiian Ridge cannot be equated with features like the Mid-Atlantic Ridge. Consequently, the rates of sedimentation over vast areas of the North Pacific are exceedingly low and growth of nodular deposits is extensive over vast areas.

Based on data on the worldwide distribution of ferromanganese, the nodules are considered to occur most frequently and over greater areas in the deep waters of the North Pacific than in any other area of the world. Dredging and coring operations have again reported the deposits most often between 6° 30' N and 20° N from deep waters off Central America at 110° W to 180° W.

Inspection of the analyses of ferromanganese deposits from the red clays of the North Pacific reveals that the average values are Ni 0.76%, Cu 0.49%, Mn 18.2% and Co 0.25%. The band of siliceous radiolarian ooze and clay south of the red clays is covered with nodules which are richer in the metals of interest. Averages for Ni are 1.28%, Cu 1.16%, Mn 24.6% and Co 0.23%. Copper and nickel values are nearly twice as great for nodules obtained from the siliceous oozes suggesting that they should be of particular interest to the ocean miner. These represent the highest values of copper and nickel determined on ferromanganese deposits of the world's oceans. The siliceous oozes also consistently include approximately 6% more Mn than their counterparts taken from red clay regions.

Cobalt does not follow the trends of Ni, Cu, and Mn; its value is essentially the same for both deep-water sediment regions. Cobalt is less abundant in nodules from the areas of red clay and siliceous ooze but is enriched in deposits located on submarine elevations such as the Hawaiian Islands, Necker Ridge and Johnston Island. Analyses indicate the nodules from red clay have 0.25% Co, those from the siliceous ooze 0.23% and values leap to an average of 0.79% on seamounts associated with or part of the Hawaiian Islands and surrounding seamounts.

South Pacific Ocean

Density of core data from the South Pacific is only half that of the North Pacific. Based on the 550 cores, there is a region of high nodule frequency lying between 10°S and 19°S and extending from 134°W to 162°W. This rectangular area includes the Manihiki Plateau, Society Islands, Tahiti, and the Tuamotu Archipelago. Nodules occur on the flanks and deeps of inter-mountain basins throughout the area.

Associated sediments are detrital carbonate on submarine slopes and red clays in adjacent deeps. For example, nodules are abundant on the east flank of the Manihiki Plateau, which is the site of carbonate sand, silt and clay in water depths of 3,500 meters and in adjacent areas of red clay deposition accumulating at depths from 4,500 to 5,000 meters.

Farther south, between 40°S and 60°S, nodules are common at the tops of cores. They are most common in the deep region of abyssal hills lying between the west flank of the East Pacific Ridge and the wide submarine plateau seaward of New Zealand. The nodules are associated with slowly accumulating red clays of the Southwestern Pacific Basin.

Although there are at least two major provinces of nodules within the South Pacific, great distances between the sites and potential markets, along with associated logistical problems, present serious obstacles to economic recovery of these ferromanganese deposits.

The framework of the South Pacific is similar to the North Pacific. However, fewer samples of ferromanganese have been recovered and less is known about the sediments of the region. Nodules and crusts are commonly encountered over submarine plateaus, ridges and seamounts (e.g., north-central South Pacific). They have been most frequently encountered in the regions of the Manihiki Plateau; Line, Cook and Society Islands; and the Tuamotu Archipelago. Nodules also occur in deep water at the flanks of the East Pacific Ridge and are scattered throughout areas of glacial erratics south of 40°S.

The South Pacific receives very little sediment from continental sources, far less than other oceans. Nodular deposits in brown clays are reported in the literature. However, based on the data most ferromanganese is located as stated above, on or in the vicinity of submarine highs such as the Line Islands, Manihiki Plateau, Cook Islands, Society Islands and Tuamotu Plateau.

Inspection of the maps reveals that, although ferromanganese concretions are very common here, the metal contents are lower than those required by the ocean miner. In regions of red clay such as the Peru Basin and deep waters east of the Marquesas Islands and Tuamotu Archipelago, however, there is a suggestion that nodules contain average values of Ni which may be attractive to the mining community in the future.

Based on available information, the South Pacific is not as attractive as the North Pacific. Average values for metals contained in ferromanganese deposits in deep-water clay regions of the South Pacific are Cu 0.23%, Ni 0.51%, Mn 15.1% and Co 0.34%. Those deposits on submarine elevations have values of Cu 0.13%, Ni 0.41%, Mn 14.6% and Co 0.78%.

North Atlantic Ocean

The distribution of ferromanganese deposits in the North Atlantic is restricted. Although this region has been sampled as extensively as any in the world, only scattered occurrences have been reported. This is especially true of the northeast Atlantic. The explanation for the paucity of nodules is the relatively high rate of deposition which prevails throughout much of the North Atlantic.

Sedimentation is dominated by the influx of immense quantities of land-derived sediment and pelagic carbonate detritus. As a result most of the floor of the North Atlantic is receiving sediment at rates in excess of those which permit development of ferromanganese precipitates.

Encrusting deposits are encountered on topographic highs such as the Kelvin Seamounts and as extensive pavements, slabs and nodules on the current-swept Blake Plateau. Another region of nodules lies east of Florida and Cuba in waters 4,500 to 5,000 m deep. This is a red clay province protected from sediment influx from either land or the Mid-Atlantic Ridge.

Samples of ferromanganese have been recovered from three different situations: the Blake Plateau; a red clay region 1,100 miles east of Florida; and submarine elevations (Kelvin Seamounts and Mid-Atlantic Ridge). These areas represent very little or no net accumulation of sediment at the site. The Blake Plateau appears to offer very little in the way of value from the point of view of economic recovery: the metals of interest occur in very low amounts in the nodules and crusts (averages Cu 0.08%, Ni 0.50%, Mn 14.5%, Co 0.42%, and there is the additional problem of high incorporated carbonated in the deposits.

Topographic barriers and great water depth have resulted in a protected region of the seabed 1,100 miles east of Florida. This area, referred to as the red clay province for convenience, contains many nodules. However, as is true of most nodules from areas of red clay, these have Ni and Cu values far below those needed for commercial recovery. The average values are Cu 0.24%, Ni 0.36%, Mn 13.9% and Co 0.35%. The encrusting deposits of the flanks, shoulders and summits of submarine elevations are equally low in the metals of interest (averages Cu 0.14%, Ni 0.39%, Mn 13.5% and Co 0.36%).

Ferromanganese is generally associated with rugged topographic features and rock exposures. Low metal values plus hazardous conditions for dredging operations suggest that it is highly unlikely that these deposits will be of interest to the mining industry.

The situation in the Atlantic Ocean is considerably different from that of the Pacific. The Atlantic is only one-third as wide as the Pacific. Major rivers deliver enormous volumes of continental debris directly to the continental margin. This material is then widely distributed by turbidity current activity to deepest parts of the ocean. There are few major obstacles to seaward dispersal of sediment and vast abyssal plains have been built up through rapid accumulation of these deposits.

In addition, there is the Mid-Atlantic Ridge System which occupies approximately half the area of the ocean. The mountains of the ridge generally lie above the compensation depth of calcium carbonate (4,200 to 4,600 m). Therefore, it is a site of relatively rapid accumulation of carbonate-rich biogenic oozes. The pelagic organic sedimentation of the ridge and terrigenous deposition on the abyssal plains occurs at rates in excess of those which permit the growth of manganese nodules. Potential nuclei of manganese precipitation are removed from the water-sediment interface by burial before accretion of manganese can take place.

Much of the North Atlantic is barren of nodules because of prevailing rapid rates of deposition. There are local concentrations of nodules and crusts in current-swept topographic highs such as the Blake Plateau and isolated seamounts (Kelvin Seamount Group) as previously indicated.

The one area of the deep ocean floor in the southwest corner of the North Atlantic as noted may be a potential economic source of metals. This area is an isolated region of red clay which lies beyond the influences of both terrigenous and biogenic contributions. Here the depths are great with nodules being recovered at stations with depths between 4,471 and 5,869 m (average 5,325 m).

Because the nodules occur in a region close to markets in the United States and Europe, more data should be gathered from this area to determine the limits, metal content, chemistry and other properties of seafloor materials.

South Atlantic and Indian Oceans

The South Atlantic and Indian Oceans are very similar to the North Atlantic in that their framework of sedimentation is dominated by large-scale input of continental debris and significant contributions from a mid-ocean ridge system. Prevailing rates of sedimentation throughout much of these oceans are above those which favor ferromanganese accretion.

As a result, the metal-rich deposits are confined to specific areas of slow deposition in protected basins or on topographic highs. Because the content of Ni and Cu in the ferromanganese deposits is very low, no attempt is made to differentiate them. Although of great interest to researchers concerned with genesis of the sediments, the crusts and nodules of the South Atlantic and Indian Oceans do not appear to offer much to the ocean miner. When the values of the various metals are averaged for the two oceans, the values found are Ni 0.54%, Cu 0.20%, Mn 16.28% and Co 0.26%. These numbers reveal the poor value of the deposits.

Nodules in the South Atlantic and neighboring areas of the Indian Ocean have developed within a common physiographic setting and regime of sedimentation. The situation is in general the same as in the southwest corner of the North Atlantic. Nodules occur most frequently within broad zones between the flanks of major mid-ocean ridge systems and abyssal plains immediately seaward of continental margins.

Several lines of evidence suggest that the contrast between manganese deposits on the eastern and western sides of the Atlantic Ocean is influenced primarily by the free flow of bottom water in the western part. There are several lines of evidence which indicate a strong flow of bottom water in the southeastern part of the Atlantic. The concentration of nodules south and west of the tip of Africa is probably related to this strong flow, which is confined to the Cape Basin by the Walvis Ridge.

In the North Atlantic, South Atlantic and Indian Oceans, nodules were found most frequently in cores taken from areas of very slow deposition between abyssal plains and mid-ocean ridge systems. This is the reason why areas of maximum nodule recovery lie within zones generally parallel to the coastlines of South America and Africa.

Ferromanganese deposits of the South Atlantic and western Indian Oceans are restricted to areas of either nondeposition or negligible sediment accumulation. Ferromanganese is more widespread in the South Atlantic than the North Atlantic. The South American Province of nodules and crusts lies between an area of rapid deposition of terrigenous sediment and carbonate-rich deposits of the Mid-Atlantic Ridge. The water is 4,500 m deep and the prevailing sediment is very fine-grained, light gray clay.

Many nodules have been recovered from the deep Cape Basin (5,000 m) lying off the tip of South Africa. The red and pink clays of the basin indicate nondeposition or very low rates of sediment accumulation. The Agulhas Plateau lies due south of the tip of Africa. Ferromanganese deposits occur on the crest of the plateaus (3,000 m) and in contiguous deeps (5,000 m).

It is quite likely that the Cape Basin Province and the Agulhas Plateau are parts of the same ferromanganese region, the Agulhas Plateau simply being the most extensive topographic high within the province. In the western Indian Ocean many cores have recovered nodules from the Madagascar and Crozet Basins. It appears that these deep basins are protected from significant influx of sediment and that conditions necessary for nodule growth prevail.

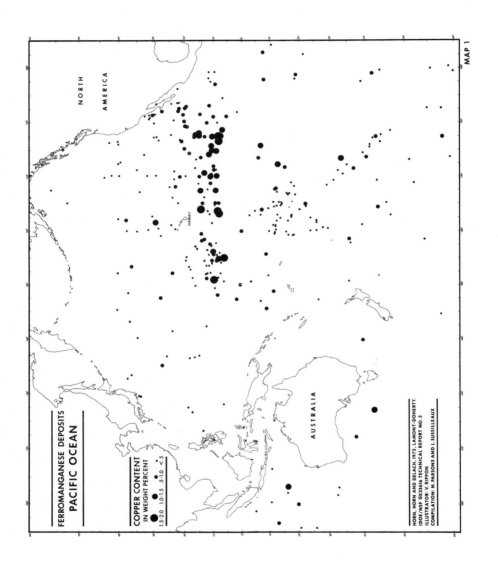

FERROMANGANESE DEPOSITS
PACIFIC OCEAN

COPPER CONTENT
IN WEIGHT PERCENT
1.5-2.0 1.0-1.5 .5-1.0 <.5

NORTH AMERICA

HAWAII

AUSTRALIA

HORN, HORN AND DELACH, 1972, LAMONT-DOHERTY.
IDOE/NSF GX33616 TECHNICAL REPORT NO. 3
ILLUSTRATOR: V. RIPPON
COMPILATION: M. PARSONS AND L. SUSSILLEAUX

MAP 1

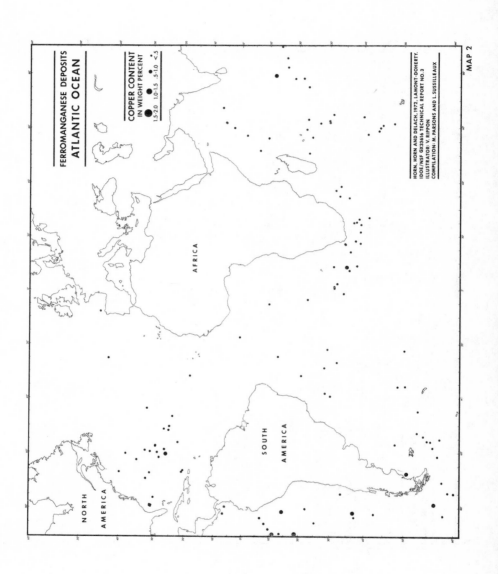

FERROMANGANESE DEPOSITS
ATLANTIC OCEAN

COPPER CONTENT
IN WEIGHT PERCENT
1.5-2.0   1.0-1.5   .5-1.0   <.5

NORTH
AMERICA

AFRICA

SOUTH
AMERICA

HORN, HORN AND DELACH, 1972, LAMONT-DOHERTY.
IDOE/NSF GX28146 TECHNICAL REPORT NO. 3
ILLUSTRATOR: V. RIPPON
COMPILATION: M. PARSONS AND L. SUSSILLEAUX

MAP 2

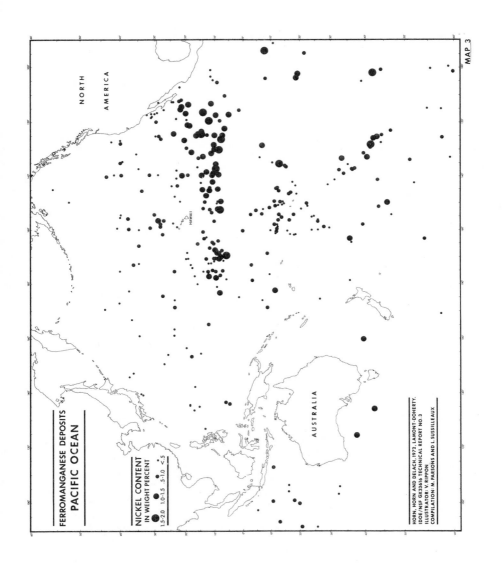

FERROMANGANESE DEPOSITS
PACIFIC OCEAN

NICKEL CONTENT
IN WEIGHT PERCENT

1.5-2.0   1.0-1.5   .5-1.0   <.5

NORTH
AMERICA

HAWAII

AUSTRALIA

HORN, HORN AND DELACH, 1972. LAMONT-DOHERTY.
IDOE/NSF GX33616 TECHNICAL REPORT NO. 3
ILLUSTRATOR: V. RIPPON
COMPILATION: M. PARSONS AND I. SUSSILLEAUX

MAP 3

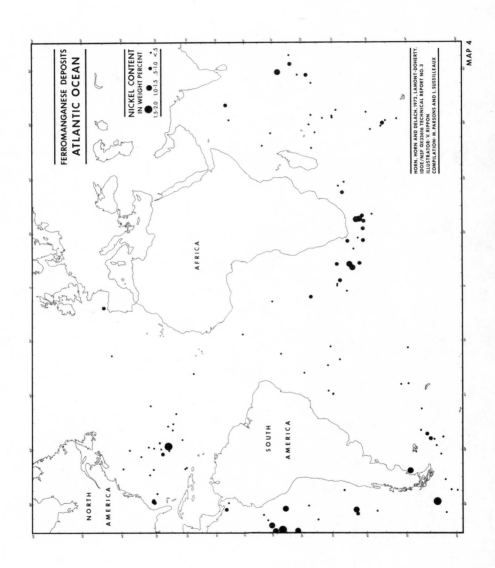

FERROMANGANESE DEPOSITS
ATLANTIC OCEAN

NICKEL CONTENT
IN WEIGHT PERCENT

1.5-2.0   1.0-1.5   .5-1.0   <.5

AFRICA

NORTH
AMERICA

SOUTH
AMERICA

HORN, HORN AND DELACH, 1972, LAMONT-DOHERTY.
IDOE/NSF GX33616 TECHNICAL REPORT NO.3
ILLUSTRATOR: V. RIPPON
COMPILATION: M. PARSONS AND L. SUSSILLEAUX

MAP 4

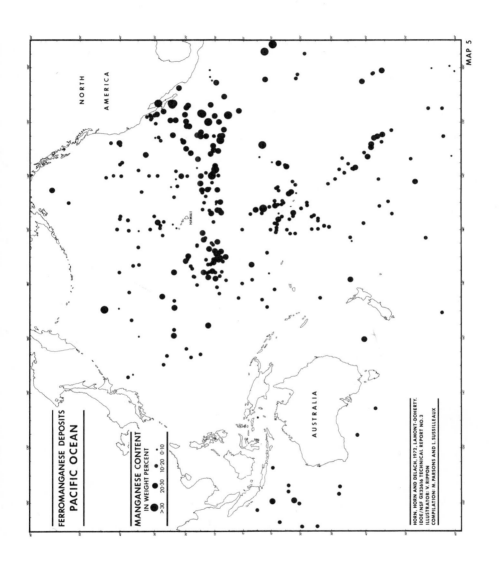

FERROMANGANESE DEPOSITS
PACIFIC OCEAN

MANGANESE CONTENT
IN WEIGHT PERCENT
>30 20-30 10-20 0-10

HORN, HORN AND DELACH, 1972. LAMONT-DOHERTY.
IDOE/NSF GX33616 TECHNICAL REPORT NO.3
ILLUSTRATOR: V. RIPPON
COMPILATION: M. PARSONS AND L. SUSSILLEAUX

MAP 5

NORTH
AMERICA

HAWAII

AUSTRALIA

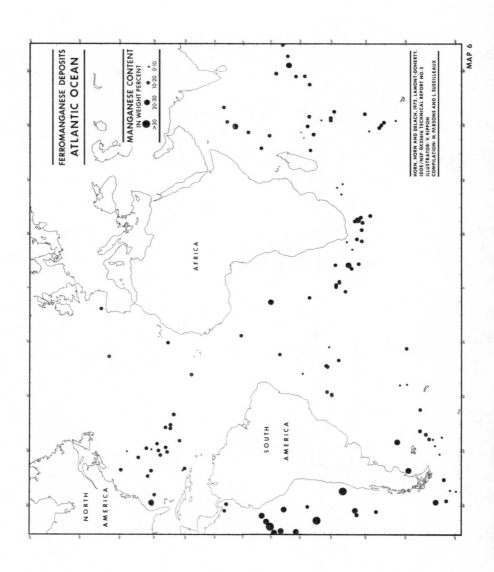

FERROMANGANESE DEPOSITS
ATLANTIC OCEAN

MANGANESE CONTENT
IN WEIGHT PERCENT

● ● ● ·
>30 20-30 10-20 0-10

NORTH
AMERICA

AFRICA

SOUTH
AMERICA

HORN, HORN AND DELACH, 1972, LAMONT-DOHERTY.
IDOE/NSF GX33616 TECHNICAL REPORT NO.3
ILLUSTRATOR: V. KIPPON
COMPILATION: M. PAXSONS AND I. SUSSILLEAUX

MAP 6

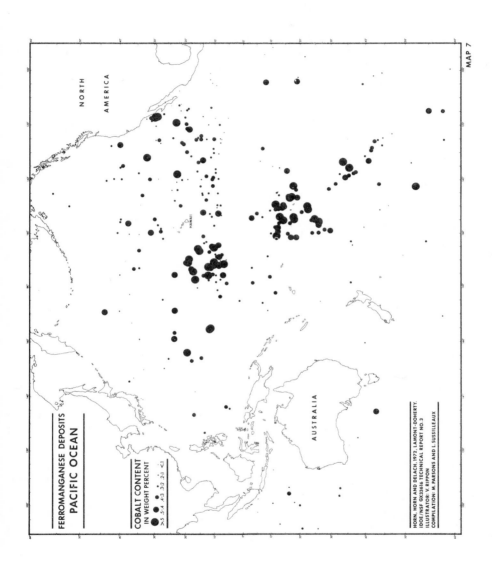

FERROMANGANESE DEPOSITS
PACIFIC OCEAN

COBALT CONTENT
IN WEIGHT PERCENT
>.5 .5-.4 .4-.3 .3-.2 .2-.1 <.1

HORN, HORN AND DELACH, 1972, LAMONT-DOHERTY.
IDOE/NSF GX33616 TECHNICAL REPORT NO. 3
ILLUSTRATOR: V. RIPPON
COMPILATION: M. PARSONS AND L. SUSSILLEAUX

MAP 7

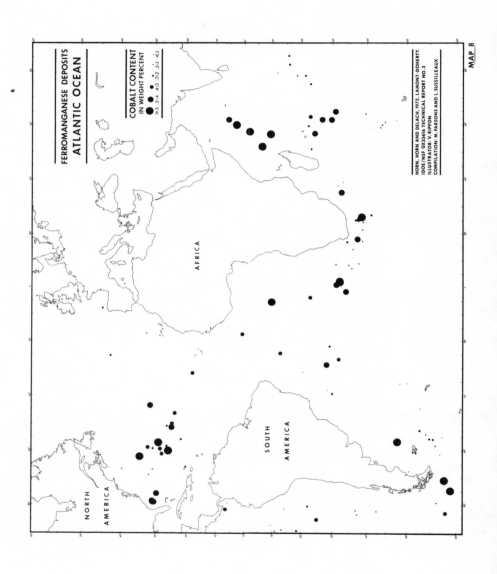

FERROMANGANESE DEPOSITS
ATLANTIC OCEAN

COBALT CONTENT
IN WEIGHT PERCENT

>5 5-4 4-3 3-2 2-1 <1

AFRICA

SOUTH
AMERICA

NORTH
AMERICA

HORN, HORN AND DELACH, 1972, LAMONT-DOHERTY.
IDOE/NSF GX33616 TECHNICAL REPORT NO. 3
ILLUSTRATOR: V. RIPPON
COMPILATION: M. PARSONS AND L. SUSSILLEAUX

MAP 8

Conclusions

Inspection of the worldwide distribution of ferromanganese deposits reveals that they are most abundant within the North Pacific (Maps 1-8, PB 223,083). They lie in an east-west band between 6° 30' N and 20° N extending from 110° W to 180° W. Not only are they most widespread here, they contain the highest values of copper and nickel of any nodular deposit in the world ocean. This region, therefore, offers the most promise to ocean miners.

The restricted distribution of the Cu-Ni rich deposits to a specific area of the North Pacific is best explained by nondeposition or minimal sedimentation in this area for millions of years. Tertiary exposures and subcrops lie immediately below the sediment-water interface. Such conditions are excellent for the development of ferromanganese deposits and possibly through time permit concentration of the metals of interest to the ocean mining community.

Major provinces of nodules occur where rates of sedimentation are lowest. Outer limits of regions of high nodule concentration mark increases in sedimentation rate due to terrigenous or biologic contributions. In the Pacific Ocean the ridges exert less control on the bottom water circulation and nodules are most frequent within east-west provinces lying north and south of the biogenic oozes of the Equatorial Pacific. They are most commonly associated with red clays. In the Atlantic and Indian Oceans, sedimentation rates are considerably higher along continental margins and over mid-ocean ridge systems.

However, there are zones between abyssal plains and mid-ocean ridges where sedimentation rates are very low. Red clays are the principal sediment. The configuration of these provinces is parallel to the continental margin on one side and to the mid-ocean ridge on the other. Volcanic debris, generally of sand- and granule-sized basaltic glass and palagonite, is the most common material around which manganese is precipitated.

Within the Northern Hemisphere, core data suggest that the east-west zone lying between 20° N and 6° 30' N in the North Pacific, and the smaller area in deep waters of the southwest corner of the North Atlantic are the most promising sites for future exploration for nodules.

REGIONAL GEOCHEMISTRY OF FERROMANGANESE NODULES IN THE WORLD OCEANS

The following information is taken from a report presented by D.S. Cronan at the conference on Ferromanganese Deposits on the Ocean Floor held at Arden House, Harriman, New York and Lamont-Doherty Geological Observatory, Columbia University, Palisades, New York, January 20-22, 1972 and issued as *PB 225 986*.

Regional variations in the composition of manganese nodules occur in the Pacific, Indian and Atlantic Oceans. In both the Pacific and Indian Oceans the variations are similar. Manganese, nickel and copper attain maximum concentrations in deposits from the east of each ocean and decrease in concentration toward the west, while the converse is found for Fe, Co, Ti, and Pb. Exceptions include the Mexican continental borderland where deposits are rich in manganese, but low in nickel and copper; and elevated submarine volcanic areas which, irrespective of their location, seem to contain deposits similar in composition. Regional variations in nodule composition in the Atlantic seem less distinct. Both manganese and iron are most abundant in separate areas of the South Atlantic, and vary fairly irregularly in the North Atlantic. Nickel, cobalt, copper, and zinc follow manganese.

On a worldwide basis, regional variations in nodule composition can probably be related to the proximity of the deposits to potential sources of metals and to the nature of their environment of deposition. One of the most important aspects of the latter is the redox potential. This influences: the degree of diagenetic mobility of manganese and hence the Mn/Fe ratio of the deposits; and their mineralogy and thus their minor element content. Redox potentials probably vary with depth, distance from land, and sedimentation rates; and thus each of these factors is likely to influence the regional geochemistry of submarine manganese deposits.

Pacific Ocean

The maximum, minimum and average weight percentages on a dry weight basis as determined by x-ray emission spectrography of 27 elements in manganese deposits from the Pacific Ocean are given in Table 1.1.

TABLE 1.1

	- - - Pacific Ocean — Statistics on 54 Samples - - -		
Element	Maximum	Minimum	Average
B	0.06	0.007	0.029
Na	4.7	1.5	2.6
Mg	2.4	1.0	1.7
Al	6.9	0.8	2.9
Si	20.1	1.3	9.4
K	3.1	0.3	0.8
Ca	4.4	0.8	1.9
Sc	0.003	0.001	0.001
Ti	1.7	0.11	0.67
V	0.11	0.21	0.054
Cr	0.007	0.001	0.001
Mn	50.9	8.2	24.2
Fe	26.6	2.4	14.0
Co	2.3	0.014	0.35
Ni	2.0	0.16	0.99
Cu	1.6	0.028	0.53
Zn	0.08	0.04	0.047
Ga	0.008	0.0002	0.001
Sr	0.16	0.024	0.081
Y	0.045	0.033	0.016
Zr	0.12	0.009	0.063
Mo	0.15	0.01	0.002
Ag	0.0006	-	0.0003*
Ba	0.64	0.08	0.18

(continued)

TABLE 1.1 (continued)

	- - - Pacific Ocean — Statistics on 54 Samples - - -		
Element	Maximum	Minimum	Average
La	0.024	0.009	0.016
Yb	0.0066	0.0013	0.0031
Pb	0.36	0.02	0.09
LOI**	39.0	15.5	25.8

*Average of 5 samples in which Ag was detected.
**LOI = Loss on ignition at 1100°F for one hour. The LOI figures are based on a total weight of air-dried sample basis.

Source: U.S. Patent 3,169,856

The average compositions of nodules from different areas of the Pacific Ocean are given in Table 1.2.

TABLE 1.2: AVERAGE COMPOSITION* OF SURFACE NODULES FROM DIFFERENT AREAS WITHIN THE PACIFIC OCEAN

	1	2	3	4	5	6	7	8	9
Mn	15.85	33.98	22.33	19.81	15.71	16.61	16.87	13.96	12.29
Fe	12.22	1.62	9.44	10.20	9.06	13.92	13.30	13.10	12.00
Ni	0.348	0.097	1.080	0.961	0.956	0.433	0.564	0.393	0.422
Co	0.514	0.0075	0.192	0.164	0.213	0.595	0.395	1.127	0.144
Cu	0.077	0.065	0.627	0.311	0.711	0.185	0.393	0.061	0.294
Pb	0.085	0.006	0.028	0.030	0.049	0.073	0.034	0.174	0.015
Ba	0.306	0.171	0.381	0.145	0.155	0.230	0.152	0.274	0.196
Mo	0.040	0.072	0.047	0.037	0.041	0.035	0.037	0.042	0.018
V	0.065	0.031	0.041	0.031	0.036	0.050	0.044	0.054	0.037
Cr	0.0051	0.0019	0.0007	0.0005	0.0012	0.0007	0.0007	0.0011	0.0044
Ti	0.489	0.060	0.425	0.467	0.561	1.007	0.810	0.773	0.634
L.O.I.	24.78	21.96	24.75	27.21	22.12	28.73	25.50	30.87	22.52
Depth (m)	1146	3003	4537	4324	5049	3539	5001	1757	4990

1) Southern Borderland Smt. Province
2) Continental borderland off Baja California
3) Northeast Pacific
4) Southeast Pacific
5) Central Pacific
6) South Pacific
7) West Pacific
8) Mid-Pacific Mountains
9) North Pacific

*Weight percent.

Source: PB 225 986

Manganese is high, greater than 20%, in the pelagic areas of low sedimentation rates in the northeast tropical and and southeast Pacific, and decreases in a westerly direction. It is intermediate in concentration, 15 to 20%, in the equatorial zone of high sedimentation rates, and reaches generally intermediate to low concentrations in nodules from the West and northwest Pacific. Minimum values of less than 5% occur in the vicinity of the Phillippine Trench and off the coast of Central America. The highest manganese concentrations, greater than 30%, occur in nodules from some marginal areas such as the American continental borderland off Baja California and southeast of the Galapagos Islands.

Iron behaves in a reverse manner to manganese. It is low in the East Pacific with values averaging between 5 and 10%, but is slightly higher in the equatorial zone. The lowest values of all 1 to 2%, occur in the continental borderland areas where manganese is enriched. It increases in a westerly direction across the Pacific reaching maximum values in the vicinity of some of the Island groups in the South and West Pacific, in the region of the Phillippine Trench and on the Macquarie Ridge. High values also occur in the manganese-low areas of Central America.

The contrasting behavior of manganese and iron in Pacific nodules is well illustrated by variations in the Mn/Fe ratio. Maximum values occur in the continental borderland areas underlain by reducing sediments, are high in the northeast tropical Pacific area of low sedimentation rates, and decrease in a westerly direction to values of near or less than unity in parts of the southern and western Pacific. Superimposed on this general trend, however, are nodules from seamounts and volcanic mountain areas. These tend to have a Mn/Fe ratio slightly in excess of 1 irrespective of their location relative to the margins of the ocean.

To a certain extent the minor elements in Pacific nodules follow the major elements, but there are some exceptions. In general, Ni and Cu follow manganese and are enriched in nodules containing the mineral todorokite from the pelagic areas of the north tropical and East Pacific. Both elements are high in the pelagic northeast Pacific, but Cu is reduced relative to Ni in the southeast Pacific (Table 1.2).

Both elements generally decrease in sedimentation. Low values of each also occur in the vicinity of the Island groups of the South Pacific, and Cu is especially low in nodules from seamounts. The coherence of Ni and Cu with Mn breaks down in the manganese-rich nodules from continental borderland areas. Here low values of each of these elements are common.

Cobalt and lead vary in a reverse manner to nickel and copper. The maximum Co and Pb values occur in birnessite-rich nodules (some of which are enriched in iron) from the elevated central volcanic areas of the oceans such as the Mid-Pacific Mountains and the Island groups of the South Pacific; but they are low in the Fe-rich nodules from some peripheral areas of the Ocean.

Other elements show less distinct variations in Pacific nodules than do those mentioned so far, but definite trends still exist. Zinc and molybdenum tend to follow Ni, Cu and Mn, in being enriched in the pelagic regions of the East Pacific, whereas, excluding perhaps the peripheral areas of the ocean, Ti and V tend to follow Fe. The coherence between Mo and Mn is so close that Mo is also enriched in the Mn-rich continental borderland nodules suggesting that it too might undergo diagenetic remobilization and reprecipitation. Ti and Sn show a similar behavior to each other with high concentrations of each occurring in nodules associated with Island groups in the South Pacific and in those from the West Pacific.

Chromium is an exception to most elements in nodules in that its distribution is not related to either their iron or their manganese phases. It is concentrated largely in unweathered detrital silicates and its distribution in Pacific concretions more or less parallels that of silicate detritus. It reaches quite high concentrations in some nodules from volcanic areas which contain a relatively high proportion of only partially weathered volcanic material.

Two suites of elements in nodules from the southern portion of the South Pacific have been delineated. These are a manganese related suite consisting of Ni, Mo, Zn, V, Cu, Co, Ba, Sn and Sr, and an Fe related suite consisting of Ti, V, Zn, Zr, Mo, Co, Ba and Sr. The former occurs in the southwest Pacific basin to the east of New Zealand, and the latter on the Albatross Cordillera (Pacific Antarctic Ridge). Both suites occur on the Chile Rise—Albatross Cordillera. The Mn suite is associated with a pelagic basin and the Fe suite with an elevated volcanic area, similar to the associations of each of these elements elsewhere in the Pacific. Regional variations in the composition of Pacific nodules are summarized in Figure 1.1.

FIGURE 1.1: SALIENT CHEMICAL CHARACTERISTICS OF NODULES AND ENCRUSTATIONS FROM DIFFERENT REGIONS OF THE PACIFIC OCEAN

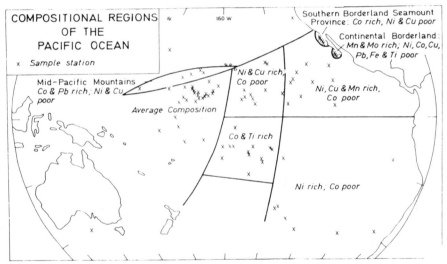

Source: PB 225 986

Indian Ocean

Regional variations in nodule composition in the Indian Ocean are less clear than in the Pacific, as there are fewer samples available on which to base an assessment. However, those samples that have been analyzed indicate that regional geochemical variations in Indian Ocean nodules are generally similar to those occurring in the Pacific. They are summarized in Figure 1.2.

FIGURE 1.2: SALIENT CHEMICAL CHARACTERISTICS OF NODULES AND ENCRUSTATIONS FROM THE EASTERN AND WESTERN INDIAN OCEAN

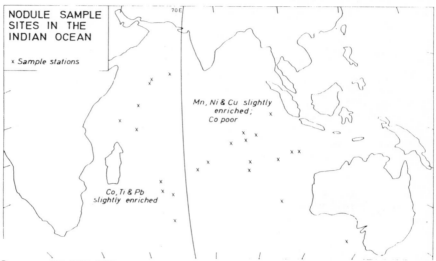

Source: PB 225 986

Manganese is highest in the east where it reaches concentrations up to 29%, and low in the west where values as low as 12% were obtained. The highest Mn values occur close to the eastern margin of the ocean and the lowest on the volcanic Carlsberg Ridge in the northwestern Indian Ocean. Iron behaves the reverse of manganese, being highest in the west, especially on the Carlsberg Ridge, and lowest in the east. Within the western part of the Ocean, however, low iron values also occur in depressions such as the Mascarene and Kerguelen Basins between the elevated volcanic areas.

Of the minor elements, Ni and Cu are highest in the east; but in the western Indian Ocean relief seems to play a large part in determining their distribution. They are generally low in the elevated areas such as the Carlsberg Ridge and the Mascarene Ridge, especially copper, but are often high in the intervening basins. In this respect their behavior is similar to that occurring in parts of the Pacific.

Also, as in the Pacific, Co and Pb vary in a reverse manner to Ni and Cu. They are highest in the elevated areas of the western Indian Ocean, and are lower in the depressed areas and over most of the eastern Indian Ocean. In certain cases, content of other minor elements in Indian Ocean nodules also show differences between the eastern and the western basins. Titanium and vanadium are highest in the western basin, whereas Mo is highest in the east. The average composition of nodules from the western and eastern basins, respectively, is presented in Table 1.3.

TABLE 1.3: AVERAGE COMPOSITION* OF SURFACE NODULES FROM DIFFERENT REGIONS OF THE INDIAN OCEAN

	West Indian Ocean	East Indian Ocean
Mn	13.56	15.83
Fe	15.75	11.31
Ni	0.322	0.512
Co	0.358	0.153
Cu	0.102	0.330
Pb	0.061	0.034
Ba	0.146	0.155
Mo	0.029	0.031
V	0.051	0.040
Cr	0.0020	0.0009
Ti	0.820	0.582
LOI	25.89	27.18
Depth (m)	3,793	5,046

Note: Geographic regions as shown in Figure 1.2.
*Weight percent.

Source: PB 225 986

Atlantic Ocean

Overall variations in the composition of Atlantic nodules are probably at least as large as those in the Pacific and Indian Oceans. However, regional variations in their composition seem less distinct than in the other two oceans.

The distribution of manganese in Atlantic concretions is shown in Figure 1.3. It is most abundant in the central and eastern South Atlantic, where values of up to 30% and more occur, and lowest in the area between South America and Antarctica. Over much of Tropical Atlantic, Mn values are between 10 and 15%. In the North Atlantic there is a broad band of values between 15 and 25% extending southeast toward the Mid-Atlantic Ridge from the southeastern United States, and then extending northeast towards the Iberian Peninsula and northwestern Europe.

In the central North Atlantic, most values fall between 8 and 15%. This distribution seems more complex than that in the Pacific and Indian Oceans where the major variations in Mn content of the concretions occur from east to west across the basin. The maximum, minimum, and average weight percentages on a dry weight basis as determined by x-ray emission spectrography of 27 elements in manganese deposits from the Atlantic Ocean are given in Table 1.4.

The distribution of iron in the concretions is shown in Figure 1.4. As would be expected, it is largely the reverse of that of manganese. The highest values occur in the southwestern Atlantic, up to 40% or more on a detrital-free basis. Over much of the South Atlantic, however, the Fe values are less than 20%, but increase above this value in the tropical area. In the North Atlantic, the distribution of Fe is irregular. However, it tends to be lowest in two areas in mid-latitudes, and varies between 20 and 30% over much of the remainder of the basin.

FIGURE 1.3: TENTATIVE DISTRIBUTION OF MANGANESE IN NODULES AND ENCRUSTATIONS FROM THE ATLANTIC OCEAN

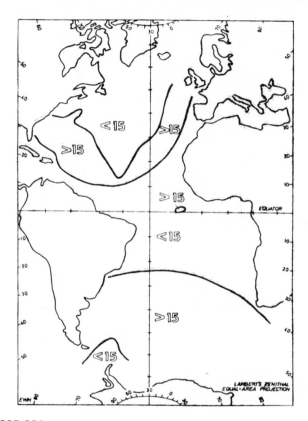

Source: PB 225 986

TABLE 1.4

Element	- - - Atlantic Ocean—Statistics on 4 Samples - - -		
	Maximum	Minimum	Average
B	0.05	0.009	0.03
Na	3.5	1.4	2.3
Mg	2.4	1.4	1.7
Al	5.8	1.4	3.1
Si	19.6	2.8	11.0
K	0.8	0.6	0.7
Ca	3.4	1.5	2.7
Sc	0.003	0.002	0.002
Ti	1.3	0.3	0.8
V	0.11	0.02	0.07
Cr	0.003	0.001	0.002
Mn	21.5	12.0	16.3
Fe	25.9	9.1	17.5

(continued)

TABLE 1.4 (continued)

| Element | Atlantic Ocean—Statistics on 4 Samples | | |
	Maximum	Minimum	Average
Co	0.68	0.06	0.31
Ni	0.54	0.31	0.42
Cu	0.41	0.05	0.20
Zn	--	--	--
Ga	--	--	--
Sr	0.14	0.04	0.09
Y	0.024	0.008	0.018
Zr	0.064	0.044	0.054
Mo	0.056	0.013	0.035
Ag	--	--	--
Ba	0.36	0.10	0.17
La	--	--	--
Yb	0.007	0.002	0.004
Pb	0.14	0.08	0.10
LOI*	30.0	17.5	23.8

*LOI = Loss on ignition at 1100°F for one hour. The LOI
figures are based on a total weight of air-dried sample basis.

Source: U.S. Patent 3,169,856

FIGURE 1.4: TENTATIVE DISTRIBUTION OF IRON IN NODULES AND ENCRUSTATIONS FROM THE ATLANTIC OCEAN

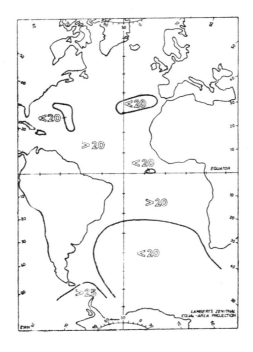

Source: PB 225 986

In general, Fe occurs in more than average and Mn in less than average concentrations in Mid-Atlantic Ridge samples. Compositional variations in Mn and Fe associated with the Mid-Atlantic Ridge in the South Atlantic might be expected on the basis of the behavior of these elements in the North Atlantic. However, no such variations are indicated in Figures 1.3 and 1.4 as no samples from the area in question were available for analysis.

Of the minor elements, Ni, Co, Cu and Zn follow manganese. In this respect Ni, Cu and Zn are behaving in a manner similar to their behavior in Pacific nodules but Co is not. Sodium, potassium, calcium and magnesium follow neither Fe nor Mn, but both Na and K, and Mg and Ca, vary with each other.

DISTRIBUTION OF MANGANESE NODULES IN THE HAWAIIAN ARCHIPELAGO

The following information was obtained from *PB 218 948*.

The geographic location of the State of Hawaii is most favorable in regard to future mining of ferromanganese deposits at abyssal depths. Based on oceanographic data, a broad band of nodular ferromanganese, rich in copper and nickel, lies 500 miles south of the Hawaiian Islands. A compilation of chemical analyses reveals the nodules have higher nickel and copper values over greater areas than in other regions of the oceans. Hawaii is the closest land to many of the deposits and is located mid-way between major markets in Japan and the United States. Therefore, it would seem that Hawaii will play a role in mining this valuable ocean resource.

A factor favoring Hawaii as a center for manganese nodule work is its Foreign Trade Zone, part of the Department of Planning and Economic Development. It is one of seven such zones in the United States. Material may be landed and processed, and the original or processed products shipped to other countries without payment of import duties. The Trade Zone has other special attractions and incentives for international commerce.

During 1970, the manganese research group at the University of Hawaii first recognized some of the interesting occurrences and associations of ferromanganese deposits at relatively shallow depths in the Hawaiian archipelago.

The deposits are accreting principally on the three prominent terrace levels around all of the Islands — at 400 to 800 meters, 1,200 to 1,600 meters, and 2,400 meters as crusts and pavements. The sediments on which they are forming are volcanogenic sands derived from Island weathering, and transported to the terraces via turbidity currents. There appears to be a distinct association between the ferromanganese crusts, their chemistry and mineralogy, and the sediments of the substrate.

The thickest and richest material appears to be accreting where the following environmental criteria are met: good supplies of iron-rich volcanogenic sediments (sands and silts) from the Islands; terraces to help trap and hold these sediments; and exposure to the flow of currents around the Islands – particularly where they are accelerated through channels between the Islands, or over shallow peaks.

The importance of the current patterns should be emphasized. Any topographic feature which obstructs a current flow will increase the current velocity in its vicinity by virtue of the diversion. This in turn exposes the substrate to a great volume of water per unit of time, and at the same time maintains the exposure of growing oxide crusts by inhibiting sediment deposition. Only the rapid large-scale deposition of turbidities would interrupt this pattern, but this would in turn provide fresh iron-rich seed material to continue the growth processes with a new pavement.

The effects of more rapid surface currents and the rich seed supply from the Islands are in marked contrast to the deep-sea environment where bottom currents move more slowly, and pelagic sediments routinely accumulate to help bury developing deposits. Only small areas of structural influence such as fracture zones promote acceleration of the currents, and other processes must act to keep the manganese nodules of the deep-sea floor exposed to seawater. Growth rates of deep-sea nodules are generally close to 1 mm/10^6 years (range of ½ to 10 mm/10^6 years).

On the Waho Terrace—a 1,900 km^2 terrace at 1,600-meter depth which extends northwest from Kaena Point, Oahu into the Kauai channel—sediments dated by hydration-rind techniques average 750,000 years old. These sediments normally have one or more crusts of 1 to 2 centimeters, or more, in thickness. This suggests rapid accretion rates of several centimeters, per million years—another facet of the shallow-depth formation of ferromanganese oxides.

Other examples of the rapid growth of ferromanganese oxides and the importance of seed material come from nodules growing on artificial seeds. A large number of iron and steel samples encrusted with ferromanganese deposits have been collected by divers during the past year off Oahu beaches. A variety of knife blades, old nails, and bottle caps with deposits may be rather old. However, ceramic-jacketed spark plugs are more recent and several of these serving as seeds for nodules have been found.

One nodule was about 5 centimeters in diameter, with over ½ centimeter of ferromanganese oxide—a growth rate several orders of magnitude above deep-sea rates, and a subject for continued research. In these artificial nodules it is interesting to note the cementing action of iron oxides migrating away from the seed—securing trapped sediment and extending the seed for manganese growth.

The German research vessel R/V *Valdivia* has provided some high-resolution echo sounding records which point up some of the structures of the terraces under examination, as well as showing the importance of narrow-beam sonar systems for detailed mapping of any future ocean mine sites.

INDIAN/ANTARCTIC NODULES

The following information was obtained from *PB 218 948*.

Most of the nodules are basically concentrated in the region of low sedimentation associated with the Southwestern Pacific Basin. Of particular interest to New Zealand is the distribution of manganese nodules over the Manihiki Plateau. Basically, these deposits occur in a depth range of 2,000 to 5,000 meters, their surface concentration within the area is strongly related to the bottom topography and they occur in concentrations of up to 50 to 70 kg/m² of sediment surface. Over one ton of nodules was obtained from a single dredge haul in this particular area and the petrography of these deposits indicates that they represent thick deposits of manganese overlaying small volcanic cores.

Since nodules from this area are cobalt-rich, these deposits may constitute an economically viable deposit. This is, of course, of interest to New Zealand as they lie off the Cook Islands which are internally self-governing, but which come under the jurisdiction of the 1964 New Zealand Continental Shelf Act. Within the New Zealand region itself, there are limited resources of manganese deposits. On the Campbell Plateau, thin encrustations of manganese overlying tertiary phosphorite deposits are reported.

The distribution of nodules across the Indian/Antarctic Ridge System is a region of very abundant nodule deposits, and has been named the Tasman Manganese Pavement. The distribution of manganese deposits at two stations taken approximately 500 kilometers north of the Indian/Antarctic Ridge System were studied.

TABLE 1.5: CHEMICAL ANALYSIS OF MANGANESE DEPOSITS FROM THE INDIAN ANTARCTIC RIDGE*

Sample No.	% Fe_2O_3	% MnO	Co	Ni	Cu	Zn
Z 2139A**	17.67	20.90	4,494	2,578	585	196
Z 2139B**	15.57	23.31	2,886	1,942	385	191
Z 2139C**	19.26	19.94	3,172	8,404	422	191
Z 2139D**	16.42	18.76	1,286	3,560	605	90
Z 2140B**	13.73	9.65	1,621	2,440	2,155	190
Z 2140 Botryoidal nodule sample 1***	11.35	22.35	2,231	6,411	1,181	162
Z 2140 Botryoidal nodule sample 2***	11.05	18.72	1,190	5,123	1,487	188
Z 2140 Small rounded nodule***	11.47	19.70	1,003	1,021	1,955	164
Z 2140 Large rounded nodule sample 1***	14.31	17.38	2,238	6,294	2,127	189
Z 2140 Large rounded nodule sample 2***	18.36	12.06	668	3,353	1,435	184
Z 2140 Large rounded nodule sample 3***	15.56	25.85	1,136	5,305	2,835	183
Z 2140 Large rounded nodule sample 4***	11.91	21.53	1,155	2,045	2,070	183
Z 2140 Large rounded nodule sample 5***	13.00	22.30	1,698	8,649	2,185	183
Z 2140 Pyramidal nodule***	16.21	13.41	1,240	3,056	1,542	183
Mean	14.71	18.99	1,858	4,299	1,498	177
Percent Standard Deviation	18.6	23.9	56.6	56.5	51.6	15.3
Percent Analytical Precision	14.41	4.65	16.02	9.67	10.39	3.90

*Analyses in ppm except where otherwise stated.
**Glacial erratic nucleus.
***Volcanic nucleus.

Source: PB 218 948

Table 1.5 shows the composition of the nodules for these localities. To facilitate analysis, the manganese oxide coating was scraped off the deposit, taking great care not to incorporate the nodule nucleus into the sample. The resultant powders were then analyzed by atomic absorption spectrophotometry. Because of the care in sample preparation and the uniform appearance of the samples, uniform analytical data were anticipated. This was not in fact the case.

Iron varies between 11 and 19% as Fe_2O_3, manganese varies between 10 and 26% as MnO, cobalt between 670 and 4,500 (parts per million), nickel between 1,000 and 8,000 ppm, copper between 380 and 2,800 ppm, and zinc between 90 and 190 ppm. Manganese and iron show a variation of a factor of 2, and cobalt, nickel and copper of a factor of 7 to 8.5.

These data emphasize the variability in nodule composition (particularly in the ore metals: copper, nickel and cobalt) over a localized area on the sea floor. This observation is particularly relevant if nodules are to be regarded as an ore resource since a detailed knowledge of the statistics of sampling of nodules on the ocean floor will be required in any mining operation.

CHEMICAL MAPPING OF THE OCEAN FLOOR—A RAPID ANALYTICAL SYSTEM

This information was derived from a report presented by J. Greenslate, R.W. Fitzgerald, J.Z. Frazer and G. Arrhenius at a conference on Ferromanganese Deposits on the Ocean Floor held at Arden House, Harriman, New York and Lamont-Doherty Geological Observatory, Columbia University, Palisades, N.Y., January 20–22, 1972 and issued as *PB 226 006;* and from a report prepared by J.F. Frazer and G. Arrhenius as part of the Inter-University Program of Research on Ferromanganese Deposits of the Ocean Floor and issued as *PB 234 011.*

The most important clues to potentially valuable deposits come from geochemical studies of the ferromanganese nodules themselves and their associated sediments. If optimal use is to be made of the results of these studies, it is essential that they be brought together, organized systematically, and made easily available to all interested parties.

Thus systems for storage and retrieval of geological data are of critical importance to efficient exploration and use of ocean resources. The systems must be computerized in order to handle the large bulk of data that is being generated at an increasingly rapid rate.

The vast expanse of the world ocean demands that any effort toward its characterization be based on huge quantities of data. For such large quantities of information to be interpretable it is necessary to implement electronic data handling systems. Understanding the chemistry of ocean sediment is one of the most complex problems in oceanography. Its solution most certainly requires rapid analyses on a large scale and the capability to interrelate the analytical results.

A tool which satisfies the requirements for rapid, accurate, nondestructive chemical analysis has been developed at Scripps Institution of Oceanography. X-ray energy spectrometry utilizing the principles of x-ray fluorescence has been shown to yield reliable quantitative (± 5%) results. The technique avoids problems associated with wave-length spectrometry such as extensive sample preparation, low emission intensities and excessive time required for scanning the entire wave-length range. Furthermore it has been demonstrated that in most cases reasonable semiquantitative results may be obtained from wet sediments immediately upon their removal from the sampling device. Analysis time for obtaining reliable data simultaneously on all elements of atomic number greater than 18 is on the order of 100 sec.

The data-handling requirements after analysis are met by the computerized Sediment Data Bank. At present the bank contains locations, depths, and brief qualitative descriptions of more than 38,000 stations in the world ocean. The stored data were extracted from published information on ocean sediment in the international literature and from various unpublished sources. Data can be recalled on the basis of specific parameters to the extent that they are available (e.g., sediment type, carbonate content).

Printout or plotting of data for selected geographic areas is routine. The system is being expanded to allow for storage of unrestricted additional amounts of analytical data for each sampling, and for direct transfer of chemical data from the analytical system to the data bank. Additionally correlation programs are being developed to facilitate interpretation of data in storage.

In an area of the Pacific Ocean Basin under study, about 300 sediment samples have been analyzed. The intent of the study is to explore regional variations in the composition of the sedimentary components, including the transition element precipitates which constitute the major fraction of manganese nodules.

This computerized World Ocean Sediment Data Bank was set up at Scripps Institution of Oceanography in 1966 and has been continually updated and expanded over the past five years. The bank contains data extracted both from the published literature and from unpublished core descriptions and analytical data of various government agencies and oceanographic research institutions throughout the world.

Its long-range purpose is to provide up-to-date and systematic information about the locations of stations in the deep ocean (> 200 meters water depth) where sediment or nodules have been collected, the general character of the sediment at those locations, results of geochemical analyses of sediments and nodules, and references to more complete information about each sample.

At the present time the data bank contains sample descriptions for more than 38,500 stations. The information includes geographical coordinates, ship or cruise name and station number, water depth, reference source, sampling device, and a brief description of the sediment lithology. The information is stored on magnetic tape and is coded in such a way that the data can be sorted according to various parameters or combinations of parameters and can be listed, classified, and mapped by computer.

Also stored in the sediment data bank are approximately 800 chemical analyses of ferromanganese nodules. Besides continuing to add new information of the types already stored in the data bank, analytical data on the geochemistry of sediments is being collected and will be added to the data bank and should be available to the public by 1973.

METAL VALUES AND MINING SITES

The following information is taken from *PB 223 130.*

When all reports of the occurrence of ferromanganese on the seafloor are plotted on world charts (Figure 1.5), the evidence shows that the Atlantic and Indian Oceans include poorly developed provinces of nodular ferromanganese. The deposits of the Manihiki Plateau and surrounding waters are widespread. The North Pacific, however, has by far the most extensive deposits over large areas. The densest concentration lies between 6°N and 20°N and from 110°W to 180°W. These are immense because conditions there have been ideal for the development of nodular deposits.

The reasons for the poorer development of major provinces of nodules and crusts in the Atlantic and Indian Oceans are: these regions receive considerable quantities of continental and biogenic debris; rates of sedimentation are high and preclude development of the nodules; and potential nuclei of the nodules are removed from the sediment-water interface through burial before accretion of ferromanganese can take place.

The situation in the North Pacific is different. There is little addition of either continental or biogenic debris. The sediment-water interface of the seafloor has remained exposed for millions of years. Nodule development can proceed uninterrupted for extended periods without disturbance. As a result, the region north of the equatorial biogenic belt of the North Pacific is marked by the largest development of nodules in the world ocean.

Areas Outside the North Pacific

Inspection of all data from the North and South Atlantic, Indian and South Pacific Oceans reveals they are secondary to the major deposits of the North Pacific. Many deposits beyond the North Pacific are of interest however because they lie close to major markets and they must be inspected carefully prior to discarding them as potential resource areas.

There are four regions of ferromanganese in the North Atlantic: (1) Kelvin Seamount, (2) Blake Plateau, (3) red clay province, and (4) Mid-Atlantic Ridge. The Kelvin Seamount is an ancient volcano covered by thick crusts of ferromanganese. Rugged relief, potential hang-ups of dredging equipment, rock exposures and very low metal contents (Ni 0.11%, Cu 0.04%, Co 0.04% and Mn 1.80%) render this area unsuitable for mining operations.

The Blake Plateau lies immediately off the coast of Florida in less than 1,000 meters of water. It was used as a test site for Deep Sea Ventures' mining operations. Crusts, slabs and nodules lie in carbonate sand. Ocean-floor photographs reveal large current ripples and scourmarks in the vicinity of the nodules and crusts. Although the deposits are close to shore on a shallow, level platform, the metal contents are low (Ni 0.52%, Cu 0.08%, Co 0.42% and Mn 14.7%).

There is the additional problem of the incorporation of large amounts of carbonate ($CaCO_3$ 10%) with the ferromanganese as the nodules grow. The low metal values suggest these deposits are of no immediate value to the mining industry.

FIGURE 1.5: WORLDWIDE DISTRIBUTION OF FERROMANGANESE DEPOSITS

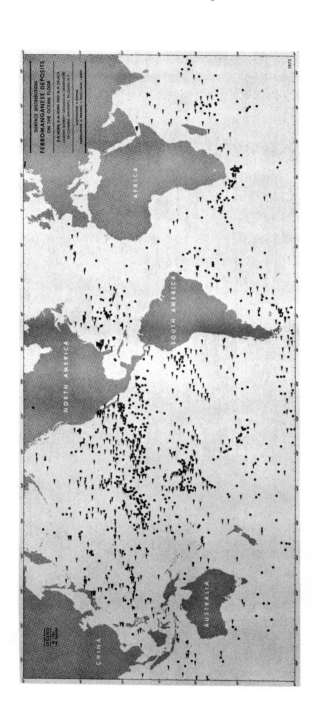

Source: PB 223 130

Within the province of red clay located 1,100 miles east of Florida, nodules are abundant. Low metal values (Ni 0.41%, Cu 0.29%, Co 0.40% and Mn 14.9%), irregular bottom, rock exposures, and patchy distribution make this an unsuitable site for ocean mining.

Finally, there are the thick crusts often encountered on the Mid-Atlantic Ridge. Rugged relief, potential hang-ups to dredging gear and low metal values (Ni 0.18%, Cu 0.12%, Co 0.3% and Mn 10.4%) make this region unsatisfactory.

All ferromanganese deposits of the North Atlantic are close to major markets in the United States and Europe, but the low metal content of the nodules eliminates the provinces as economic sources of nickel and copper.

There are fewer deposits in the South Atlantic, and they have a tendency to be irregular and patchy. There is one poorly developed locality off the east coast of South America. The Rio Grande Rise lies within the province. Thick crusts are associated with rock exposures, and nodules are widely distributed around this topographic feature. Low nodule density and low metal values (Ni 0.14%, Cu 0.09%, Co 0.05% and Mn 7.2%) make it unlikely that either the rise or the waters surrounding it offer a potential mining site.

On the other side of the Atlantic off the tip of South Africa are extensive areas of ferromanganese nodules and crusts. Many of the nodules in deep basins lie shoulder to shoulder and contain intermediate nickel and low copper values (Ni 0.67%, Cu 0.16%). On the Agulhas Plateau large crusts and nodules are encountered in relatively shallow water, but the metal contents are relatively low (Ni 0.83%, Cu 0.15%). Low to moderate nickel and poor Cu components tend to eliminate this area for commercial exploitation.

There is an abundance of nodules in the Madagascar Basin. They are larger than the average nodule and range in size from baseball to football dimensions. Rock exposures, variable continuity of coverage, and low metal values (Ni 0.24%, Cu 0.12%, Co 0.25% and Mn 11.0%) suggest the deposits are not of economic interest. The Crozet Basin in the southern Indian Ocean is littered with closely spaced nodules forming extensive fields. Low metal contents (Ni 0.42%, Cu 0.12%, Co 0.14% and Mn 12.4%) indicate they are not suitable for mining.

The large region of high concentration of ferromanganese deposits in the central South Pacific can be characterized by the Manihiki Plateau. The plateau itself is a site of carbonate deposition. Red clay prevails in surrounding deep waters. Both the crest of the plateau and nearby deep areas are blanketed by extensive concentrations of nodules.

The ferromanganese deposits of the level plateau are associated with rock exposures, carbonate sands and muds, potential hang-ups and low metal values (Ni 0.30%, Cu 0.17%). Dense concentrations of nodules occur in adjacent basins, but they too have low metal values (Ni 0.13% and Cu 0.10%).

It is concluded that low metal values of all deposits of the Atlantic, Indian and South Pacific Oceans render them of little interest to the ocean mining community.

The North Pacific

From the previous discussion it is noted that vast areas of the world ocean have major provinces of ferromanganese, yet the metal values of the deposits are so low they are of no interest ot the ocean miner. Now, if all available chemical analyses, including the North Pacific, are plotted, the results will show a definite concentration of copper and nickel within the North Pacific (Copper, Maps 1 and 2; Nickel, Maps 3 and 4). These Cu-Ni-rich deposits are the only ones of their kind found to-date in the world ocean.

Inspection of the sedimentary provinces of the North Pacific reveals that the nodules are equally abundant in the red clays and siliceous deposits. Ocean-floor photographs, however, suggest there are considerably fewer nodules associated with the red clay areas. Highest metal values are confined to a narrow band of siliceous deposits north of the equator. It is here that the most promising fields of Cu-Ni-rich deposits occur in the world ocean. To the ocean miner the nodules associated with the red clay are of little interest because they contain approximately half the amounts of nickel and copper as their counterparts from the siliceous deposits.

Good nodules from the siliceous region are pancake- to hamburger-shaped. They have a relatively smooth top and mammillate fuzzy undersurface. There is often a change in size at the nodule's equator, the lower portion being smaller. The top is often smoothed by erosion, whereas the bottom consists of a series of protuberances and hairs associated with downward growth.

If nickel in weight percent is plotted against water depth, nodules containing more than 1% nickel are associated with siliceous sediments (Figure 1.6). The nickel-rich nodules occur most frequently in depths between 4,000 and 5,600 meters, with the majority at 4,900 meters. Poor nodules from regions of red clay are generally smaller, subequant, multilobate and smooth.

Nickel-Copper-Rich Nodules of the North Pacific: Nodules which contain high values of nickel and copper lie within an east-west band in the southeast and southcentral North Pacific. The regional slope is toward the west and northwest (Figure 1.7). There is no correlation between regional slope and metal content of the nodules.

The substrate throughout the copper-nickel-rich band is siliceous ooze and clay. Framework grains of the deposits are siliceous skeletons of Radiolaria. These sediments have up to 88% porosity and this may be a factor in the concentration of nickel and copper in the nodules.

Examination of bottom photographs of the region reveals a tendency for burial of the nodules at the eastern end of the nodule zone. Burial is most likely due to an influx of continental and biogenic detritus from Central America and the East Pacific Rise. Whatever the cause, the eastern nodules are partially to totally buried by a thin veneer of sediment, whereas those in the central and western parts of the band are free of sediment cover.

It has been stated that most ferromanganese nodules occur at the sediment-water interface. A plot of the information from the North Pacific supports this hypothesis.

FIGURE 1.6: NICKEL CONTENT

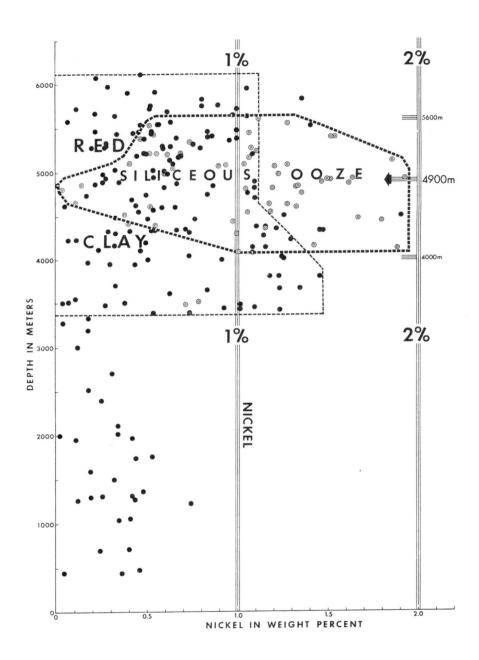

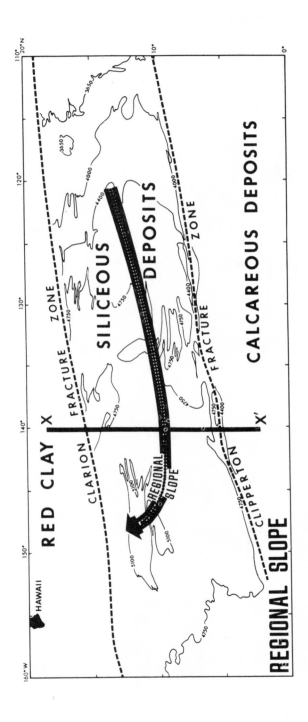

FIGURE 1.7: REGIONAL SLOPE CONTAINING HIGH NICKEL AND COPPER VALUES

Source: PB 223 130

FIGURE 1.8: NODULE OCCURRENCE AT SEDIMENT-WATER INTERFACI

Source: PB 223 130

FIGURE 1.9: EAST-WEST AXIS OF MAXIMUM COPPER CONTENT

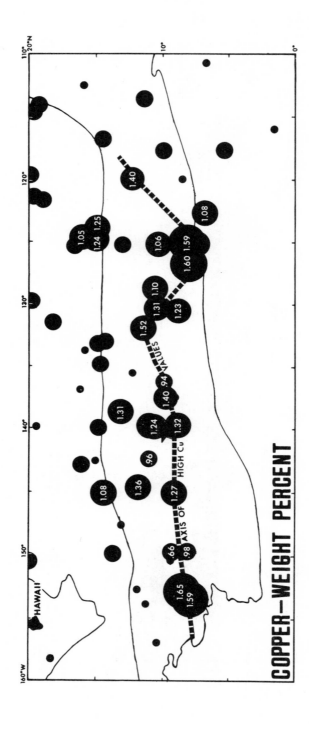

FIGURE 1.10: EAST-WEST AXIS OF MAXIMUM NICKEL CONTENT

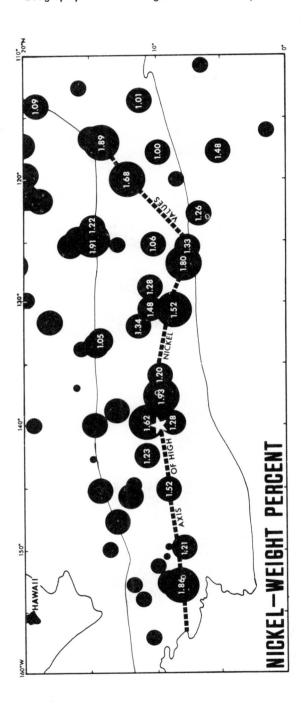

NICKEL—WEIGHT PERCENT

Source: PB 223 130

FIGURE 1.11: MINING SITES

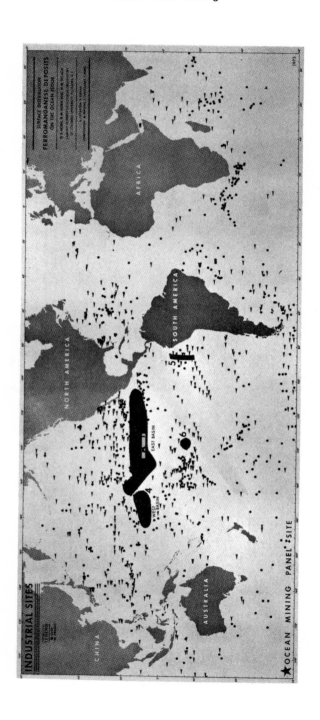

The majority rest on the surface of the sediment (Figure 1.8), which indicates there is no need to penetrate the substrate during dredging operations. If the values of copper are plotted within the siliceous band there is a definite east-west axis (Figure 1.9) of maximum copper content. The same is true of the nickel distribution (Figure 1.10). When both are compared, the maximum copper and nickel duplicate one another. If these axes are overprinted on a map of sediment age, it is clear that the maximum values of copper and nickel are found in nodules recovered from Miocene sediment.

Mining Sites

There is no question that the best area, for both economic and research interests, lies within the North Pacific band of siliceous deposits. Here, almost a continuous carpet of nodules lies at the surface. Topography consists of gently rolling hills. There is a wide range in age of the sediments on which the nodules rest.

Submarine erosion appears to have been active in the region for extended periods of time. The nodules are large and provide ample evidence as to their genesis. This site is equally important to the mining industry. Here the most lucrative deposits rich in nickel and copper occur.

The Ocean Mining Panel of National Security Industrial Associates has selected a site for extensive study within the band at 10°N and 140°W. Gently rolling topography and an abundance of surficial nodules should insure successful recovery of ferromanganese at the site.

The most promising area in the ocean for mining sites is the east-west belt in the southeast corner of the North Pacific (Figure 1.11). Greatest potential lies in mining sites located along a narrow band defined by 8°30' N 150°W to 10°N 131°30'W.

TECHNOLOGY RELATING TO MANGANESE NODULE RECOVERY

DEEP SEA NODULE MINING SYSTEMS

The following is taken from a paper presented by J.L. Mero at the conference on Ferromanganese Deposits on the Ocean Floor, held at Arden House, Harriman, New York and Lamont-Doherty Geological Observatory, Columbia University, Palisades, New York, January 20-22, 1972 and issued as *PB 226,004.*

Although numerous systems have been conceived for the recovery of nodules from the ocean floor, only two appear to have merit from an economic standpoint: the hydraulic system and the cable line bucket system.

The hydraulic system generally consists of a length of pipe which is suspended from a surface float or vessel; a gathering head, designed to collect and winnow the nodules from the surface sediments and feed them to the bottom of the pipeline while rejecting oversized material; and some means of causing the water inside the pipeline to flow upward with sufficient velocity to suck the nodules into the system and transport them to the surface.

The two power means for hydraulic dredges being considered are conventional centrifugal dredge pumps and air lift pumps. In 1970, one company successfully tested an air lift dredge in the Blake Plateau nodule deposit in about 2,500 ft (760 m) of water. Generally, the hydraulic systems are quite complicated and, thus, expensive.

Capital investments in systems being planned or under construction range from about $30 to $60 million for systems capable of recovering about one million tons of nodules per year from depths as great as 18,000 ft (5,490 m) of water. The estimated operating costs of these systems range from about $10 to $20 per ton of nodules produced at the surface of the ocean. Two of these systems are illustrated in Figure 2.1. The second type of system for full-scale production of the nodules is a mechanical, Continuous Line Bucket (CLB) system which consists essentially of a loop of cable to which are attached dredge buckets at 25 to 50 m intervals and a traction machine on the surface vessel capable of

44

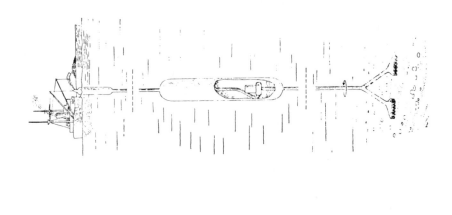

FIGURE 2.1: TWO HYDRAULIC SYSTEMS

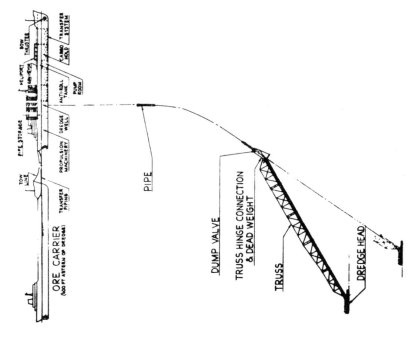

Source: PB 226,004

moving the cable such that the buckets descend to the ocean floor along one side of the loop, skim over the bottom filling with nodules along the bottom side of the loop and return to the surface on the third side of the loop. This system of dredging is illustrated in Figure 2.2.

Thus far this system of recovering the nodules has been successfully proven in a series of tests which culminated in a test in over 12,000 ft (3,650 m) of water in a deposit of the nodules lying about 250 miles north of Tahiti. Because of the great simplicity of the CLB system, problems in operating it are relatively minor and the capital costs are relatively low, in the order of $2.5 million for a system, exclusive of the cost of a surface vessel, which can recover about three million tons of nodules per year from any depth of water.

FIGURE 2.2: CONTINUOUS LINE BUCKET SYSTEM

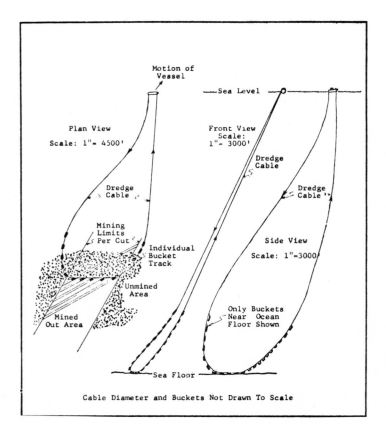

Source: PB 226,004

The estimated operating costs of this system, including the cost of chartering a surface vessel, are about $0.50 per ton of nodules recovered. The system can be mounted on practically any type of vessel capable of transiting the open ocean and of carrying a total load of about 1,000 tons. In addition to being extremely simple, the CLB system incorporates a very high degree of flexibility in being able to work in deposits of nodules of any size range, over relatively great sea floor topographic relief and in a range of sediment bearing strengths and characteristics.

A given CLB system can be easily modified to operate in any depth of water and all parts are surfaced several times a day for inspection and repair. Also, all complicated components of the system are located aboard the surface vessel for ease of inspection and repair. No measurable pollution of the ocean will be created by the use of the Continuous Line Bucket system as it is possible to skim the nodules from the sea floor without disturbing the sediments to any degree.

TECHNIQUES FOR NODULE RECOVERY

The following is taken from a paper presented by B.G. Stechler and J.T. Nicholas at a conference on Ferromanganese Deposits on the Ocean Floor, held at Arden House, Harriman, New York and Lamont-Doherty Geological Observatory, Columbia University, Palisades, New York, January 20-22, 1972, and issued as *PB 226, 000*.

Three possible mining methods are proposed. System A (Figure 2.3) uses a self-propelled dredge ship which tows a dredge head along the ocean floor at the end of a long suction pipe.

FIGURE 2.3: SYSTEM A

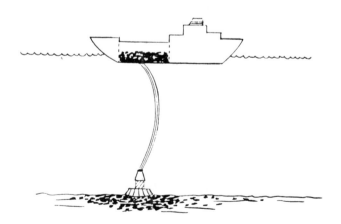

Source: PB 226,000

Dragged at 2 to 3 mph the dredge head uses its front and back teeth to select and hold on to nodules that measure 2 to 3 inches in diameter, the best size for pumping and handling. A stream of compressed air piped to the dredge head forces a slurry of nodules, silt and water into the suction pipe and, aided by hydrostatic pressure of the ocean, spits the mixture aboard ship.

This first method appears to have the advantage of a fast, simple, continuous method of recovery. The appearance, as is often the case, is misleading. Some of the conceivable problems inherent in this method are listed below.

> Numerous problems relating to the limitations imposed on the system by the hydraulics involved, e.g., dealing with multiphase flow (solid, liquid, air) at depths of 10,000 to 20,000 ft.
>
> Structural problems presented by launching and retrieving two miles of piping from a mother ship.
>
> Control of the pickup device and orientation of the suction pipe.
>
> Reliability, one break in the line stops the continuous cycle for an indefinite period of time, such a breakage would probably require replacement of the line.

System B (Figure 2.4) is a continuous loop of nylon rope 2.4 times the water depth with 7 inch buckets attached every 82 feet. Working from a 132 foot research vessel, the buckets descend from the bow and return at the stern. A pulling mechanism mounted near the stern provides the heft for lifting the mud filled buckets after they make a traverse along the bottom.

FIGURE 2.4: SYSTEM B

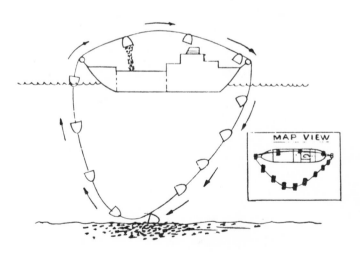

Source: PB 226,000

This method also seems quite simple. There are, however, several difficulties to consider which are listed below.

The effect of currents on the line make it difficult to home in on the exact length of line for mining.

The effect of the ship's vector velocity in the X and Y planes, plus the tension vector in the line, does not make it possible to achieve an optimum contact angle for filling the bucket.

The tension distribution along the up side of the line can cause slack on the down side part of the line. This, too, can upset the contact angle and the location of the bucket resulting in unoptimized fill for the bucket.

The nylon line is subject to abrasion, cannot exceed allowable stresses, and has drag properties that limit the rate of travel.

In System C (Figure 2.5) one mother ship with a series of 10 deep ocean mining vehicles is operated in addition to one standby unit. The cycle of operation of each mining vehicle is divided into four phases:

Phase 1 Descent to the ocean floor.
Phase 2 Harvest maneuver, payload recovery and lift off.
Phase 3 Ascent to the surface.
Phase 4 Unloading of vehicle and refitting for repeat cycle.

As in both Systems A and B, System C has some problems to overcome. By comparing Systems A and B with System C one clear difference becomes apparent, Systems A and B are tethered, System C utilizes the ocean environment to perform its objectives.

FIGURE 2.5: SYSTEM C

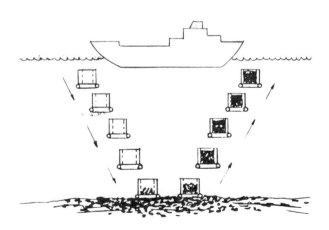

Source: PB 226,000

PROPRIETARY MINING METHODS

System Which Includes a Movable Position-Indicating Anchor

R. Vogt; U.S. Patent 3,365,823; January 30, 1968; assigned to Scientia Corp.
describes an ocean floor mining system and method which includes a movable
position-indicating anchor which is interconnected with a dredge vessel by a
guide cable, and a dredge bucket which is guided along the guide cable and which
is also interconnected with the dredge vessel by an operating cable.

With the anchor in a first position, the operating cable can pull the dredge bucket
along the ocean floor in engagement with the guide cable to a second position,
and thence up to the vessel along the guide cable, where the bucket can engage
abutment means and dump its contents. The anchor can be moved to the second position and the dredge bucket returned to the ocean floor along the guide
cable until it contacts the anchor at the second position. The bucket can then
be pulled to a third position and the cycle repeated.

Harvesting System Incorporating Continuous Aggregate Transfer Method

A.J. Nelson; U.S. Patent 3,429,062; February 25, 1969 describes a deep water
harvesting system which provides a harvester for gathering and size classification
of aggregate disbursed on the floor of a body of water, and a transfer means designed to be unhampered by turbulent action of a disturbed sea surface for recovery of the aggregate to a receptacle.

The harvester is buoyantly supported with monitoring and control means to automatically adjust its position responsive to depth and undulations in the floor for
a maintained degree of penetration in the floor. Furthermore, the harvester elements are articulately connected to flex in response to irregularities for contiguous
contact with the floor.

The transfer means is varied: when all gathered material is sought then simple
application of a pump suction provides the feed for the transfer; when undesirable fines would be entrained, then a secondary classifier is employed propelling
the accumulated material for a sufficient confining trajectory through the open
water body.

The heavier sought aggregate is collected for feed to a jet pump with clear pres-
sured water secured from a source free of contamination. The uncollected fines
fall free into the trail of the advancing harvester. Crushers are employed to recover material that is rejected because of excessive size. Means are provided to
assure level of the apparatus and, in the event of power failure, discharge of the
piping to prevent back flow into the pumping apparatus.

Apparatus Incorporating Crushing and Cleansing Means

*N. Koot and R.W. Nold; U.S. Patent 3,433,531; March 18, 1969; assigned to
Global Marine, Inc.* provide an undersea mining apparatus and method which
harvests a relatively large area of the ocean floor without moving the apparatus
thus avoiding the necessity of constantly stationing a support vessel or providing
a complex ore conduit.

The process increases the percentage of ore-bearing material raised to the surface by cleansing the ore of silt and sediment at the ocean's floor. Moreover, the chances of ore conduit clogging are reduced and the economy of undersea mining is increased by reducing the average size of ore transported to the surface.

Briefly, the apparatus comprises a bearing plate upon which a hub is rotatably mounted. At least one substantially horizontal arm is attached to the hub. Means are also provided for rotating the arm and hub as a unit about the latter's longitudinal axis. Ore harvesting means are associated with the arm to gather the ore in the path of rotation of the arm for its transfer to the ocean's surface.

A preferred form of the mining apparatus includes means for cleansing the harvested ore of sedimentary material which is often encrusted on its surface before the ore is transported to the surface. In addition, means are also provided for crushing the ore in preparation for its lift from the floor of the ocean to the surface. The crushing and cleansing means are powered by the means for rotating the arm.

The apparatus would typically be used after a preliminary survey has indicated that the floor of the ocean is relatively smooth and sufficiently extensive in ore deposits to warrant mining. The area is then harvested by simply moving the apparatus, after the segment of the floor within its reach has been mined, from place to place. In other words, all that is required for economic use of the apparatus and process is a preliminary survey which shows the extent of the undersea terrain in which commercially attractive nodules are present.

Mining is accomplished by lowering the apparatus to the ocean's floor and sweeping a given area, harvesting the ore in the path of the arm's rotation, and transporting the harvested ore to the surface. Through the use of this relatively permanent apparatus, ore can be mined without the constant necessity for moving the surface vessel in relation to the mining apparatus to compensate for ore conduit stress produced by relative movement between the surface vessel in relation to the mining apparatus. This relative permanency would, of course, overcome the necessity for a complex ore conduit design which would otherwise be necessary to avoid the stress problem.

The removal of nonproductive materials before transporting the ore to the surface is accomplished by the mining apparatus' cleansing means. Moreover, the crushing of nodules on the ocean's floor reduces the chances of clogging the ore conduit by avoiding the lodging of nodules in the conduit. In addition, the amount of power required to raise a given tonnage of ore to the ocean's surface is reduced by grinding the nodules and enhancing the efficiency of the air lift which would normally be used with the mining apparatus.

Self-Propelled Tractor Controlled by Surface Ship

J.R. Graham and A.A. Mabson; U.S. Patent 3,456,371; July 22, 1969; assigned to Kennecott Copper Corporation describes a process and apparatus for mining deposits on the sea floor such as nodules of manganese and other minerals. The sea floor is skimmed and the skimmed material mixed with seawater to serve as a carrier which conveys the mixture through a dredge pipe to the surface of the sea where the solid material is separated from the seawater.

A ship is on the surface and a self-propelled tractor operates on the sea floor and is supplied with power and controlled from the ship. This equipment, utilizing air-lifting techniques for causing transport of the mined material, operates in the following fashion.

The surface vessel preferably is equipped with auxiliary equipment and supplies, to permit operations at a substantial distance from land bases, without the need of direct or continuous land based support other than for transportation of mined materials, food and supplies, and emergency assistance.

When an area of the sea floor that is to be mined has been established, a series of at least four taut line buoys is set to a known and recorded pattern. Each of these buoys is equipped with a sonar transponder, battery powered, with a signaling life of several months. These initial buoys form the master buoys for the area to be mined, and subsequent mapping and setting of auxiliary marker buoys, of which there will normally be very many, are related to the original master buoys. The sonar transponders are serviced and repowered, at regular intervals as needed, by divers.

After an explored area has been staked out, the surface vessel is moved into operating position. The gathering vehicle or tractor is carried by the surface vessel, either on deck in a center well or suspended directly beneath the center well, when operations are pending.

To begin operations, the tractor is slowly lowered by adding on the appropriate dredge pipe sections to lengthen the dredge pipe sufficiently to lower the tractor to the sea floor. A pendent wire line is held out from the bow of the ship on a boom as the tractor is lowered and this automatically forms the desired catenary at the lower end of the dredge pipe, as the tractor is lowered to the sea floor.

When the final section of dredge pipe has been connected the entire weight of the dredge pipe is transferred to a spider that is mounted at the level of the main deck of the surface vessel. After the dredge pipe has been secured to the spider, a receiving tank is placed in position and secured in place for operations.

A swing joint through which the dredge pipe is connected with the tractor, permits the tractor to move in wide sweeps transversely of the direction of travel of the surface vessel. This decreases the required speed of the surface vessel, and results in smaller drag forces on the dredge pipe.

The movement of the tractor on the surface of the sea floor is remotely controlled from the surface vessel, in order to cause the tractor to move in a predetermined pattern of movement, or as indicated by the topography of the sea floor, to skim the surface of the sea floor in the areas traversed. Before the tractor is started on its sweeps over the sea floor, the air lift is placed in operation to start a current of seawater up through the dredge pipe.

As the tractor moves over the sea floor, the sea floor is raked to loosen up the material forming the sea floor, and to free up the manganese nodules. Scraper blades plow all of the nodules and other solid material into a bed in the path of a collector head. This bed is immediately traversed by the collector head. As the collector head passes through the bed of nodules and other material from the sea floor, the current of seawater passing into the collector head sweeps the bed up, through the screen in the lower face of the collector head, and then through a stub pipe and the successive fittings into the dredge pipe. For efficient mining operations, the area to be mined should be relatively flat. When the nodule concentration, of nodules lying on the surface and within the first six inches of depth of the surface of the sea floor, is about four pounds per square foot of sea floor surface or greater, strip mining in accordance with this procedure ordinarily is effective and efficient.

The tractor can be driven at a speed of about one or two feet per second, and the rate of recovery of nodules may be on the order of 100 to 200 lb/sec, depending upon the size of the collector head. For mining manganese nodules at two typical operational depths, the conditions given in Table 2.1 are recommended for producing satisfactory results. In the table the values are approximate.

TABLE 2.1

	- - Water Depth (feet)- -	
	6,000	12,000
Inside diameter of dredge pipe (inches)	15.0	15.0
Fluid velocity of two phase flow (nodules, other solids, and seawater), in the bottom portion of dredge pipe (ft/sec)	17.1	17.1
Maximum velocity of nodules in two phase flow (ft/sec)	14.4	14.4
Minimum velocity of nodules in two phase flow (ft/sec)	7.7	7.7
Average probable velocity of nodules in three phase flow (solids including nodules, seawater, and air) (ft/sec)	73	73
Distance of air injection below surface of water (feet)	1,375	2,180
Percent of water depth to point of air injection	23	18
Injection air pressure to start system (gauge) (lb/in^2)	615	980
Injection air pressure to maintain flow after start (lb/in^2)	355	525
Volume of air ratio (cu ft/bbl water)	50	75
Volume of air quantity (cfm)	10,600	16,000
Compressor horsepower to maintain system	1,800	3,400
Compressor horsepower total on dredge	4,000	8,000
Total flow (slurry) (lb/min)	87,300	87,300
Total flow (nodules) (lb/min)	9,000	9,000

As indicated in the table, in order to start the initial upward flow of seawater through the dredge pipe, it is necessary to use air that is in excess of the static head of the water at the injection point so as to start a pumping effect from the injected air. However, after the water is in motion, the pressure of the air can be lowered considerably to a maintenance level that is still sufficient to keep the fluid in motion. At the collector head, the inward rush of seawater sweeps the nodules and other loose material that have been formed into a bed by the scraper blades into the collector head and up the dredge pipe.

At the upper end of the dredge pipe the material discharges into the chamber of the tank. This chamber is maintained at a superatmospheric pressure, e.g., about 22 psig or higher. The solids portion of the three phase mixture, that is discharged into the tank, settles rapidly. The solids are transferred through a rotary valve from the bottom of the tank into a material handling system such as a transfer conveyor that will transfer the solid materials into appropriate hoppers. The seawater is allowed to run off and return directly to the sea.

The figures for nodule recovery are based upon nodules having a largest dimension of about 8", about 0.6 sphericity, specific gravity 2.1, and density of about 131 lb/ft³. For operating at a 6,000 ft depth, the air injection site preferably is located about 1,375 ft below the surface of the sea, and at the operating injection pressure of 355 psia, the daily air rate is about 15.4 standard MCF. Preferably, the compressors withdraw air from the tank into which the dredge pipe discharges. This air is then compressed and returned to the dredge pipe at the injection site.

As a design precaution, it is preferred that the compressed air system be flexible, with multiple compressor units, with at least 100% of standby capacity, and reserve high pressure air tanks, all capable of delivering compressed air in a wide range of pressure. For mining in water depths of up to about 12,000 ft, an exemplary dredge pipe string would be constituted as shown in Table 2.2, from the sea floor up.

TABLE 2.2

Length (feet)	Inside Diameter (inches)	Wall Thickness (inch)	Material
6,000	15	½	Welded aluminum
1,500	15	⅝	Welded aluminum
6,000	15	½	Welded, high tensile steel
400	15	¾	Welded, high tensile steel

To minimize the power required for operations, it is preferred that the air injection site be located at a depth of at least 1,000 ft below the surface of the sea. When hydraulic lifting is employed rather than air lifting, centrifugal pumps replace the air injection system, and the dredge pipe discharges directly into the atmosphere instead of into a pressurized discharge chamber.

For mining operations down to about 6,000 ft of water, it is preferred to employ two pumps of 1,500 hp each, operating in series, located about 600 ft below the surface of the sea. These pumps should be constant speed, AC powered pumps, with adequate motors to meet maximum emergency power fluctuations. For mining operations at depths on the order of 12,000 ft, it is preferred to employ four pumps, operating in series and located about 1,000 ft below the surface.

In the event of damage to the dredge pipe or tractor, or after exhaustion of a particular mining area, the dredge pipe and tractor can be hauled aboard the surface vessel. In the event that a storm requires that the surface vessel abandon a particular area quickly, the dredge pipe can be disconnected from the surface vessel, but left connected to the buoy through the pendent line, and the dredge pipe can simply be laid on the sea floor for retrieval after the storm. Preferably, a sonar transponder or repeating target is mounted adjacent preselected joints, at intervals along the length of the dredge pipe, and, as well, on the gathering vehicle.

These units will respond to the ultrasonic sound of a sonar transmitter on the surface vessel, and their transmitted energy will in turn be received on the ship. By a suitable display of these signals in both plan and elevation, the configuration of the pipeline, and the location of the gathering vehicle relative to the surface vessel, are ascertained.

All of the electric power supply that is required to drive the bottom vehicle and control it, for the underwater television equipment and flood lights, and for the instruments, preferably is supplied in one composite cable. This power supply cable is attached to each section of the dredge pipe by means of a suitable quick-connecting clamp. For convenience, cable reels are mounted on the deck of the surface vessel, for stowing and reeling out of the cable. The cable preferably is prepared in lengths of approximately 6,000 ft, with watertight connectors at the joints. Amplifiers for the underwater television are inserted in these joints, as needed.

The composite cable is made up of individual wires as needed for power, lights, electronics and controls, all suitably assembled in one unit and then covered with a moisture impermeable insulation, protected by double armor, reverse lay wrapping. The end of the cable is let out of the center axle of the cable reel, and is terminated in suitable connectors that are connected up for operation after the vehicle is in operative position on the sea floor.

Bottom Crawler Under Control of a Floating Vessel

The deep sea nodule mining method devised by *J.E. Steele and G.W. Sheary; U.S. Patents 3,504,943; April 7, 1970; and 3,480,326; November 25, 1969; both assigned to Bethlehem Steel Corporation* provides a bottom crawler adapted to traverse the ocean floor under control of a floating vessel.

A suction pump on the crawler picks up water and solid material from the ocean floor through collector sleds and delivers it to a separator on the crawler. The crawler is provided with wheels having radially extending treads which penetrate the ocean floor and collect solid material between them. The solid material is jetted out of the treads by high pressure water, passed over a grizzly, and delivered to a separator for further classification.

Large solids fall to the bottom of the separator, while smaller solids and water flow out the top of the separator. Another pump in the riser pipe communicating between the crawler and floating vessel creates a fast upwardly moving stream of water entraining large solids from the bottom of the separator and transports the solids to a floating vessel.

Self-Contained Mining Vessel Capable of Assembling Equipment at Sea

J.E. Flipse, R.M. Donaldson, J.L. Stevens, Jr., J.L. Mero, J.D. Deal, Jr., N.D.

Birrell, N.E. Oresko, L.E. Spencer, A.E. Graham, A.L. Jones, I.L. Hirshman, F.S. Todd and W.A. Himes; U.S. Patent 3,522,670; August 4, 1970; assigned to Newport News Shipbuilding and Dry Dock Company describe an especially constructed seagoing vessel which stores components of an underwater mining apparatus. The assembled components of the apparatus are used for actual underwater mining at varying depth in excess of about 400 ft beneath the surface of the sea.

The ship is provided with equipment for handling and assembling individual components of the apparatus one-by-one and for lowering it from the ship to a mining operation on the sea bottom. When the ship with the apparatus assembled, depended and attached, is moved relative to the sea surface a successful mining operation of solids is effected. When necessary, the assembled components may be disassembled and recovered.

Involved in the operation is a basic combination of a ship which is adapted to move through the water over a particular mining area of the sea bottom along with a collecting device also adapted to move along the sea bottom to collect the solid manganese nodule bodies. A conduit is provided for transporting the collected solid bodies from the collecting means up to the ship whereat the collected bodies are stored. The collecting device is caused to move along the sea bottom through a particular mining area, the device being moved and guided in such a manner as to traverse or sweep through the mining area so as to gather the maximum number of nodules in an efficient manner.

The manganese nodules to be mined are solid generally spherical bodies normally ranging in size from small ones about the size of a toy marble to large ones the size of basketballs or even larger. As a practical matter, it is considered that mining of nodules from about ¾" to 6" in diameter may be the most economical and efficient arrangement, and accordingly the collecting means is provided with separation apparatus for collecting only solid bodies which are within the desired size range.

In the general combination, two fundamental concepts are envisioned. In a first arrangement, the ship moves under accurate guidance and at an accurate speed through the water with the conduit depending therefrom and with the collector being towed by the lower end portion of the conduit, the collector following the contour of the sea bottom.

In a second arrangement, means are provided on the collector itself for propelling or moving the collector along the sea bottom at a particular rate and is remotely guided. In this case the ship is maneuvered so as to remain in the proper operative relationship with respect to the collector.

In the system described, a completely self-contained ship is provided which is adapted to carry out the underwater mining operations far out at sea, the ship being designed to operate for extended periods of time with all of the equipment required for the underwater mining operation being carried by the ship and stowed on board. In addition, the ship includes storage means for storing the collected nodules at sea. In a modification, a pair of ships are employed operating in conjunction with one another, the first of the two ships providing storage means and propulsion means for towing the other ship, the other serving primarily as a support and storage means for the underwater mining equipment which is lowered from the ship to operate along the sea bottom. The first of the two ships is adapted to tow the second through the water over a particular mining area of the sea bottom, a collector being provided and adapted to move along the sea bottom to collect solid manganese nodule bodies. A conduit is provided for transporting the collected solid bodies from the collecting means up to the storage on the first of the two ships.

In each instance the ship supporting the underwater mining equipment is provided with a well portion located in a central part of the ship such that when the ship is in operative mining position, the well portion opens down into the sea and facilitates handling of the mining equipment which depends through this well portion.

The depending conduit includes a lower portion which is articulated with respect to the upper portion, the lower portion of the conduit being so interconnected with the upper portion as to permit movement thereof in a manner to accommodate for variations in the depth of the sea bottom from the surface. In other words, the collector which is connected with the lower end of the lower portion of the conduit may accordingly rise and fall and follow the contour and undulations of the sea bottom while the ship moves along through the water in a more or less steady path.

The lower portion of the conduit is also preferably provided with means for illuminating the bottom and for remotely viewing the bottom whereby personnel on the ship can observe the progress of the mining operation and can cause the collector to move in a proper path along the sea bottom for maximum efficiency of operation.

The articulated lower portion of the conduit includes a dead weight to which a truss portion is pivotally connected. The details of construction of the dead weight means and its pivotal interconnection with the truss means is shown in U.S. Patent 3,302,315.

The truss construction in each of the modifications may be substantially identical and includes a plurality of individual sections which can be selectively interconnected with one another particularly in the case where the first modification is concerned, whereas in the second modification the truss may be formed as a single unit. The reason for this difference is the fact that in the first modification, it is desired to disassemble the overall truss into a plurality of individual sections for the purpose of stowage, whereas in the second modification the entire truss unit may be stowed within the ship without disassembling the truss.

The collecting means in each modification may be substantially identical, and the details of construction of this type of collecting means are illustrated in U.S. Patent 3,305,950, the only difference being that here the collecting device includes a unit for propelling it along the sea bottom rather than being towed.

The collector device is particularly designed for movement along the sea bottom, and as mentioned above may either be towed along the bottom or may be provided with a propelling system for moving the collecting means in the desired manner. Various arrangements may be provided for facilitating movement of

the collecting means along the sea bottom such as wheels, endless tracks or run-
ners and various forms of propelling means may be utilized for moving the collecting means where this arrangement is desired.

The collector employs size sorting means in the form of spaced teeth which are
adapted to reject nodules which are not of the desired size so that only solid
bodies of a size which is considered desirable are collected. The collector also
includes a throat portion to which all of the collected bodies are directed, the
lower end of the conduit being connected with this throat portion for entraining
the collected nodules and moving them up to the ship.

A pump is provided which connects in the conduit for creating a flow of water
sufficient to entrain the nodules and to carry the nodules up to the ship. A particular feature is the fact that the pump is connected at an intermediate point in
the conduit which is disposed a substantial distance below the surface of the
sea. With this arrangement, the water and the nodules are in effect sucked upward
within the conduit to the pump, whereupon the water and the nodules are then
pumped the remaining distance to the ship.

This arrangement is considered essential in order to enable proper hydraulic performance of the pump, and the pump is preferably disposed a distance below
the surface of the sea which is approximately 10% or more of the distance from
the surface of the sea to the collecting means at the sea bottom. While the disposition of the pumping means may be varied to a certain extent, it must be
located a distance below the surface of the sea sufficient to prevent cavitation
in the pump.

The collector and the conduit as well as the pump are interrelated with one another and designed to operate at peak efficiency at a particular solids to liquid
ratio of such mixture. In order to ensure that the apparatus operates at optimum
efficiency, a remotely controlled valve is connected in the conduit and a variable
speed pump is connected in the conduit which can also be remotely controlled
to maintain a substantially constant ratio of solids to liquid of the mixture formed
through the conduit.

In order to perform the most economical mining operation, it is essential to
maintain a substantially steady course of movement of the collector through a
particular mining area. Especially in the case where the collector is being towed
special attention must be given to current conditions at the mine site. In an area
where tidal currents prevail, it will probably be necessary to set up a scheduled
ship's heading to change course with each tide change making appropriate adjust-
ments for wind and sea conditions to avoid zigzagging across the mine field.

All of this emphasizes that it is necessary to provide exacting navigational techniques to locate accurately the collector and determine its direction of travel
and movement along the sea bottom. Accordingly, navigational equipment which
provides for determining the position of the collector with respect to a fixed reference base line whereby the necessary information is obtained at all times is provided. Furthermore, in order to properly direct the mining operation, means is
provided for plotting the position and heading of the collector with respect to
such a base line. Means is also provided for determining the relative position of
the collector with respect to the ship and the linear distance therebetween so

that the ship can be properly maneuvered in order to maintain its correct operating position relative to the collector. The mining ship employed in the first
modification includes a central well portion which permits ready access to the
sea for lowering large heavy components of the underwater mining equipment
into the sea and for lifting such equipment out of the sea. This well portion is
positioned to facilitate handling of the equipment on the ship and positioning
of the ship properly for lowering operations.

The ship is also provided with stabilizers which permit the ship to be precisely
controlled and accurately maneuvered into proper position. Since the mining
operation or movement of the collecting means will nominally be at speeds less
than 5 knots, a controllable reversible pitch propeller is provided for obtaining
the necessary control. In order to assure a reasonably steady ship's heading, a
bow thruster means is installed in the bow portion of the ship to counter the
effects of wind and sea.

In a typical example, the ship should be designed for a free route speed of about
14 knots for travel between port and the mine site with cargo and wing tanks
ballasted as required to maintain a draft of about 30 ft with the underwater
mining equipment. Antiroll tank means is also incorporated in the ship to minimize roll and to provide the maximum degree of stability.

Since all of the underwater mining equipment is adapted to be carried to the
mining site by the ship, the ship must include means for stowing all of the equipment on board and for handling the equipment to move it from its stowed position into operative mining position and to return it to its stowed position when
required.

The underwater mining equipment when supported in operative mining position
is suspended down through the well portion formed in the ship, and since the
collector is travelling along the bottom of the sea while the ship is moving to a
certain degree on the surface of the sea with respect to the underwater components, it is necessary to permit a certain degree of relative movement between
the conduit and the ship. Accordingly, means is provided for supporting the
conduit from the ship so that such relative movement may occur.

The mining ship is adapted to store a substantial quantity of solid bodies of collected material within the hold which is of a particular construction so as to receive such solid bodies of material. In addition, since the storage on the ship is
limited, it is essential in order to provide an economical operation to be able to
transfer the solid bodies from the holds of the mining ship to an auxiliary cargo
ship while the mining ship is at sea and conducting substantially continuous mining operations.

Accordingly, a transfer system which permits the cargo stored within the holds
of the mining ship to be transferred to another ship on the high seas is provided.
A principle involved here is to tow a cargo ship about 600 ft astern of the mining ship and to float a suitable conduit between the two vessels. The cargo is
fluidized and pumped to the cargo ship by means of the transfer system incorporated in the mining ship itself.

Conduit Payout Apparatus

In the underwater mining method described by *M.W. Smith and C.S. Kluth; U.S. Patent 3,543,527; December 1, 1970; assigned to Westinghouse Electric Corp.* a surface vessel, to which is connected an underwater mining collector, travels through the water while connecting and paying out conduit sections. A gimbal arrangement on the vessel allows the conduit to assume a certain angle with respect to vertical as the vessel travels through the water.

A derrick on the vessel is tilted at the same angle that the conduit makes with the vertical and includes a means for gripping a conduit section to be added onto the already payed out conduit. The tilted derrick serves to receive a new conduit section, lower the section down to the gimbal arrangement for joining onto the payed out conduit, and repeats the process until the desired length of conduit is payed out. Thereafter, the surface vessel trailing the conduit and collector slows down to a predetermined speed for placing the collector on the sea bottom to commence mining operations.

Collecting Device

M.W. Smith; U.S. Patent 3,556,598; January 19, 1971; assigned to Westinghouse Electric Corporation describes a collector for subsea mining operations. The collector is generally cylindrical and rotatable around a horizontal longitudinal axis. A plurality of trays cylindrically disposed about the longitudinal axis are connected to a plurality of rings which engage the seabed and rotate the trays as the collector is pulled.

A plurality of teeth, each tooth located between respective rings at the lower portion of the collector insure that scooped up aggregates are deposited into the trays. A hopper at the front of the collector structure receives the aggregate material carried by the trays when the trays revolve to a point where the aggregates may drop into the hopper. Screens in front of and behind the revolving trays allow the ambient water moving past the collector to sweep out undersize aggregates.

The ends of the hopper are connected to a suction source and when aggregates are deposited into the hopper, a baffle directs the material toward the edges of the hopper, and collection may take place. The structure is symmetrically arranged so that it can be operable in case a twist should develop in the conduit or other means which places the structure on the sea bottom.

In brief, the collector structure includes a scoop rotatable around an axis as the collector structure is moved over a terrain such as the bed of a body of water. Terrain engagement means digging into the terrain causes the rotation of the scoop.

A blade for scraping the terrain is disposed for depositing the scraped up aggregates into the scoop as it rotates. A hopper within the volume of rotation of the scoop is provided for receiving the contents of the scoop. During underwater use, first and second screens located on either side of a portion of the scoop serve to keep aggregates, of the predetermined size range, within the scoop and to allow the ambient water to sweep away undersize aggregate such as silt and sediment. The blade may be positioned on the lower portion of a follower having adjustable weights for determining the size cut that the blade takes and to balance the shear strength of the seabed. A second blade is positioned on the upper portion of the follower.

The hopper maintains substantially the same orientation during the movement of the structure and the symmetrical nature of the structure, in conjunction with the second blade, insures proper operation even if the connector to a surface vessel gets twisted thus turning the collector over.

Foldable Collector and Streamlined Riser Conduit

A.M. Rossfelder and B.J. Thorn; U.S. Patent 3,588,174; June 28, 1971; assigned to Tetra Tech, Inc. describe a collector assembly which may be towed on the ocean floor for mining loose aggregate, the aggregates being lifted to the surface by means of a water stream which flows upwardly from the collector. The collector is foldable, and it is fully deployed only when immersed and ready for operation.

Hydraulic means is provided for gathering and conveying the nodules toward the intake to the water stream. Ground traction of the collector is provided by means of a combined tow from a riser and from a follower pendant which is rigged on the collector through sidelines similar to those used in fishing gear.

The collector assembly may also include a streamlined riser conduit so that the collector may be moved freely at a relatively high ground speed. In addition, hydraulic means is provided in the collector assembly for gathering and conveying the nodules toward the suction bin at the intake to the riser. The hydraulic means also provides an agitating action on the nodules, which serves to sort the nodules, and which also serves as a means for the dispersion of fine sediment from the nodules so that such sediment is not drawn up through the riser.

As mentioned above, ground traction for the collector assembly is provided by a combined towing action of the riser and a follower pendant. The follower pendant is rigged to the collector through sidelines which, as mentioned above, have been used in fishing gear. The pendant is supported by a controllable buoyancy package which is towed by the surface craft at a lesser depth than the collector assembly. The control of the buoyancy of this package allows for control of the lifting force exerted by the pendant on the collector, and, consequently, of the net weight of the collector on the sea floor.

Extension of the gathering arms of the collector assembly to a deployed position, when the assembly to be described is submerged, is assured by the operation of the surface craft, and an automatic pressure actuated latching device may be incorporated into the assembly in order to maintain the gathering arms locked in their deployed position whenever the collector is below a certain depth.

Moreover, any appropriate remote controlled means may be used for opening and closing the gathering arms when the collector assembly is in operation on the sea floor, so as to control the rate at which the nodules are fed to the intake of the riser. Appropriate sensing means may be incorporated into the overall assembly so that the rate at which the nodules are fed through the riser to the surface is indicated, and the remote controlled means may be operated accordingly so as to maintain a predetermined flow rate. Such sensing means, for example, may be a television camera, electrical resistivity sensors, magnetometers, and the like.

A front sled may be incorporated in the collector assembly in order to support the lower end of the riser, and the sled may be used also to support a pump which sets up hydraulic flow in the assembly to draw the nodules into the gathering arms. Further ground traction may be incorporated in the form of a powered crawler or buggy.

Vertically Floating Dredging Vessel

M.G. Krutein; U.S. Patent 3,620,572; November 16, 1971 describes a sea mining vessel in which minerals lying at the sea bottom are first conveyed from the sea bottom to the submerged portion of an elongate vessel floating substantially vertically in the sea and then conveyed vertically from the submersed portion of the vessel to the surface of the sea.

The apparatus includes an elongate vessel having an extended tank portion sealed for submersion over a major portion of its length and ballast means for shifting the center of gravity of the vessel between a transport position with the sealed tank lying substantially horizontally in the sea and a mining position with the vessel and tank floating in a vertical position.

In mining position the major portion of the sealed tank is submersed in the sea, and minerals are brought to the vessel by dredging or conveying equipment mounted on the furthest submersed portion of the vessel. The dredged matter is brought into the submersed portion of the sealed tank. There the matter is dewatered and the desired minerals conveyed up through the tank to the surface of the sea.

The dredging vessel is designed to float in a vertical position during dredging or mining operations so that with the major portion of the vessel submerged, the vessel is affected to a minimum extent by wave action adjacent the surface of the sea. This mining vessel is highly stable in vertical position thereby providing both stability and maneuverability to the actual equipment that picks up and moves the mined minerals.

The stability of the mining vessel permits ocean mining at virtually any depth desired since mining equipment extending to any depth can be attached to this vertically floating vessel. In one example, the cross-sectional area of the mining or dredging vessel is reduced or suitably shaped over the region of heavy wave action to maintain the effects of wave motion on the dredging vessel at a minimum.

With a dredge so constructed and having the mining or dredging equipment which conveys the minerals from the sea bottom to the vessel located on the submerged end of the vessel, the required length of the dredging or conveying equipment is effectively reduced by the submerged length of the vessel.

This construction provides better maneuverability for the conveying equipment due to the shorter length of mining equipment in mining operations where the distance between the lower submersed end of the vessel and the sea bottom is small.

The dredging vessel is powered to move through the sea at selectible low rates of speed whereby the dredging operation can take place during patterned slow traverses across a mineral deposit lying on the sea bottom for maximum recovery of the desired minerals.

Hydraulic conveying equipment is used to convey the minerals from the sea bottom into a portion of the dredge vessel which is located far beneath the surface of the sea but maintained at atmospheric pressure due to communication with atmosphere upwardly through the vessel. This construction permits the utilization of the difference in pressure at the bottom of the sea and at the location within the dredging vessel to convey the minerals into the vessel.

After the minerals have been passed into the vessel and separated from the conveying hydraulic fluid or water, the water can then be conveniently pumped out of the vessel. This arrangement eliminates the amount of work required to move the minerals from the sea bottom to the level of the lower submersed portion of the vessel.

The matter recovered from the sea bottom and conveyed into the dredge or vessel is separated as to desired and undesired minerals. Then the desired minerals can be conveyed vertically within the vessel to the surface of the sea and the undesired minerals exhausted from the vessel back into the sea. This arrangement reduces the amount of work that would otherwise be required for the recovery of desired minerals since it is unnecessary to convey undesired minerals the submersed length of the vessel.

Apparatus Employing Injection and Suction Nozzles

The dredging method described by *E.J. Beck, Jr.; U.S. Patent 3,646,694; March 7, 1972; assigned to U.S. Secretary of the Navy* which employs injection and suction nozzles is designed for the recovery of discrete bodies of substantial mass or high density from an aqueous floor by their forceful injection into the inlet or mouth of a vacuum device which is closely associated with the injection apparatus.

In operation a fluid medium is drawn rapidly through the vacuum device and associated vacuum equipment and entrains any object forced therein. Thus, discrete heavy objects may conveniently be lifted from the ocean floor and transported to a remote location for collection or further treatment.

The apparatus comprises an injector nozzle connected by a high pressure hose or pipe to a pumping arrangement located at a suitable point above the ocean floor. The nozzle is adapted to direct a very strong hydraulic stream, supplied by a pump, against discrete objects on the aqueous floor and force such objects into an adjacent flared suction or collection head attached to suction piping which leads to the pumping arrangement.

A discharge conduit extends upwardly from the pumping arrangement and may connect to a conveyor arrangement at a remote point or may lead into the hold of a ship. The pump supplies fluid under great pressure to the high pressure pipe which fluid is ejected through a nozzle against discrete objects on the ocean floor, such as manganese nodules or the like. At the same time, the pump also constantly draws water into a collection or suction head.

Thus, any object forced into the suction head is first dislodged from the floor without unnecessary entrainment of cohesive sediments by cross currents of turbulent stream flow. Such object is then carried upwardly through a pipe, pump and exhaust line to the conveying arrangement or ship where it may be further removed to any desired treatment area.

Apparatus Employing Dredge Nets

Y. Masuda and T. Murakami; U.S. Patent 3,672,079; June 27, 1972 describe a mechanism for mining manganese nodules from the deep sea bottom which involves a number of dredge nets tied to a long endless rope suspended from both sides of a ship. The rope falls from one side of the ship to the deep sea bottom with an apparatus provided to pull the rope to the other side of the ship, whereby manganese nodules are continuously collected by the dredge nets.

Figure 2.6 illustrates the general arrangement of the mechanism in which a mining ship **1** has a front wheel **2** at a position close to the bow and a rear wheel **3** at a position close to the stern and guide wheels **4,5,6,7,8,10** and **11** on the deck. Guide plates **12** and **13** are attached closely to the guide wheels **5** and **7**. Guide roller **14** is disposed between guide wheels **4** and **5** and guide roller **15** between guide wheels **9** and **10**.

FIGURE 2.6: APPARATUS EMPLOYING DREDGE NETS

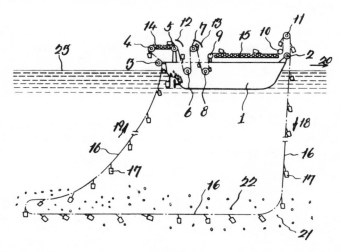

Source: Y. Masuda and T. Murakami; U.S. Patent 3,672,079; June 27, 1972

Endless rope **16** extending from the bow to the deep sea bottom by way of the stern passes through the rear wheel **3**, guide wheels **4, 5, 6, 7, 8, 9, 10, 11** and the front wheel **2**, the length of the rope being 2.4 times as long as the depth of the sea.

A number of dredge nets **17** are tied or secured in any suitable manner to the long rope at regular intervals. Guide wheels **4, 5, 6, 7, 8, 9, 10** and **11** have electrically driven motors incorporated therein. Accordingly, the long rope travels from the bow to the stern by way of the sea bottom. The arrow **18** indicates the movement direction of the forward end of the rope and arrow **19** indicates that of the rear end of the rope.

The mining ship makes headway at a slow speed. The arrow **20** indicates the direction of the ship's movement. When the speed of rope is almost equivalent to that of the ship, the forward end of the rope is suspended almost perpendicularly but the rear end thereof is pulled by the movements of the ship so that the dredge nets **17** are drawn on the clay at the deep sea bottom **21** and collect manganese nodules **22** on the clay, and are pulled up onto the sea surface **23**.

Transfer of Collected Material Through Fluid Transport System

E.P. Johnson; U.S. Patent 3,672,725; June 27, 1972; assigned to Earl and Wright describes a deep sea mining method in which an automatically operating mining vehicle works on the ocean floor in concert with a mother vessel on the ocean surface and with a material transport conveyance for the material recovered from the floor to the vessel.

All components of the mining apparatus are compatible with the deep sea environment and are adapted to operate at great depths under extremely high pressure. The vehicle is programmed to automatically traverse a predetermined path along the ocean floor while collecting, crushing and classifying the material to an optimum particle size for use in the transport system.

The crushed and sized particles are pumped through a conduit system at an optimum velocity to develop the desired material/water mixture. Discharge of the mixture from the vehicle is directed through a length of flexible conduit which is releasably connected with a relatively rigid riser conduit suspended from the mother vessel. The discharge is entrained with a stream of high pressure water pumped upwardly through the riser conduit and discharging into the vessel for processing or storage.

In one example, a pair of interconnected pipes are provided for deep water operation and water is forced through the pipes by pump means on the vessel. In another example a single riser pipe is provided for shallow water operation and the mixture is pumped through this pipe from the vehicle while an alternate boost pump may be provided to assist in lifting the mixture.

In the double pipe system discharge from the flexible conduit is entrained with the stream of water in the riser pipe by means of a venturi arrangement. The releasable connection between the flexible conduit and the riser pipe is provided by a sea socket having interfitting male and female elements with a tension cable

controlled from the vessel and adapted to lock the elements together in fluid
communication, and to release the elements permitting discharge pressure to dis-
connect the coupling. As the vehicle is raised to the vessel the cable is payed
out through the socket. When the vehicle is lowered to the ocean floor, reentry
with the socket connection is established by reeling in the cable and locking the
elements together by maintaining cable tension.

Scraper Bucket System

E.B. Dane, Jr.; U.S. Patent 3,675,348; July 11, 1972 has designed a scraper
bucket apparatus for deep sea mining systems which scrapes loose ore sediment
from a deep seabed and delivers it to an undersea mining vehicle traveling the
bed. The apparatus is embodied in a vertically flexible, long and foldable double
tiered track carrying an endless chain with a series of scraper buckets attached
that scrape the nodules and deliver them to the vehicle.

To reduce crabbing, the buckets are skewed forward of the vehicle by an angle ϕ
relative to the axis c of the track, where

$$\phi = \theta - \tan^{-1}(S_c\sin\theta)/(S_v + S_c\cos\theta)$$

where θ is the angle at which the apparatus is disposed relative to the negative
of the velocity vector of the vehicle, S_c the speed at which the chain is driven
along the track, and S_v is the speed of the vehicle. Angle θ preferably is maintained at 90° to maximize the width of the swath so that

$$\phi = 90° - \tan^{-1}S_c/S_v$$

One end of the track is rotatably mounted on a tail sheave tractor driven along
a course parallel to the vehicle and at a speed preferably to keep the track disposed at an angle of 90° relative to the negative of the vehicle's velocity vector.

Rotating Paddle Wheel Collector

*R.H. Murray; U.S. Patent 3,697,134; October 10, 1972; assigned to Bethlehem
Steel Corporation* has designed a mechanical nodule collector which is lowered
to the ocean floor and towed along by a floating towing vessel. The towing
means may be either a hoisting rope or a pipe through which collected nodules
may be continuously pumped to a floating transport vessel.

If the towing means is a pipe, the pipe is connected directly to a buoyant vessel,
or pod, which is connected to the floating towing vessel and the floating transport vessel. The collector can be readily raised or lowered by varying the buoyancy.

The collector comprises a framework having a rotating paddle wheel attached
to it which penetrates the ocean bottom and sweeps nodules up a grating where
they fall into a basket. The peripheral speed of the paddle wheel equals the
peripheral speed of treads supporting the framework so that there is little or no
relative motion between the tread and the ocean bottom, thereby minimizing
disturbance of the bottom soil during the traversal thereof by the collector.

The framework of the collector is supported on the ocean bottom by at least two treads. Rotatably attached to the framework and adapted to receive nodules is a basket. A paddle wheel having its axis concentric with the axis of rotation of the basket is rotatably connected to the framework, the inner edges of the paddle wheel blades being slightly spaced from the outer surface of the basket.

The paddle wheel is connected by gears to tread-driving means so that the peripheral speed of the paddle wheel blades is equal to the peripheral speed of the treads. This connection results in very little disturbance of the ocean bottom during the collecting operation. A grillwork, extending from a point substantially directly under the paddle wheel axis to a point higher than the axis and to the rear of it, is disposed in close proximity to the periphery of the paddle wheel. Nodules are swept up the grillwork and subsequently slide across the blades and drop into the basket.

System Using a Magnetically Nonconductive Drum

W.H. Kuhlmann-Schaefer and M.E. Dinter; U.S. Patent 3,776,593; December 4, 1973; assigned to Preussag AG, Germany describe a design for an apparatus for collecting manganese or other magnetizable material from the bottom of the sea and conveying it to the sea surface which comprises a rotating drum having teeth on its surface to lift the material from the sea bottom, a plurality of magnets mounted inside the drum extending along the axis of the drum through an arc between 90° and 180°, and a suction device mounted at the surface of the drum near the terminal ends of the magnets to carry the material from the drum to the surface.

The magnets do not rotate with the drum surface but remain disposed at the trailing edge of the rotating drum to hold the material on the surface until it reaches the suction conveyor. Preferably, the arcuate magnets are arranged side by side with the north and south poles alternating with each other in the axial direction of the drum.

The system consists of a magnetically nonconductive drum which is rotatably mounted about a horizontal spigot and rolls over the seabed, and in which there is a magnet that extends from the lowest zone of the drum in the direction of rotation over part of the drum circumference upwards and is located closely adjacent to the inner wall of the drum. The conveyor system picks up the magnetizable raw material particles carried along on the surface of the drum where the magnet ends.

This arrangement exploits the fact that lumps of manganese can be magnetized with a view to lifting only them from the seabed and passing them to the conveyor system which conveys them to the surface of the sea. Particles which cannot be magnetized are not picked up and remain on the seabed so that they do not burden the conveyor system. The magnetic forces of attraction also enable magnetizable particles which are not directly at the surface to be likewise attracted, whereby it may, in addition, be advisable to stir up the seabed to some extent. With such a lifting system stirring up may be effected directly by the lifting system.

If several lifting systems are used, it is possible to increase the depth to which stirring up takes place. The drum is guided at a distance from the seabed by radially projecting rims fitted laterally to the drum. To aid in picking up, these rims may be provided with roughening means in the form of teeth.

To avoid excessively deep penetration of the seabed or excessively firm positioning of the drum on the seabed, as a result of which the lumps of manganese may be rolled into the ground thus making it impossible to raise them magnetically, it is advisable to compensate the weight of the drum and the machinery associated with it, at least in part, by means of a buoy. In this way damage to the drum is avoided if it rolls over major rocks.

To produce the magnetic forces permanent magnets or electric magnets may be used. If electric magnets are used it is advisable to provide a supraconductive winding so that the energy losses are small and the magnetic forces generated large. The units for cooling the winding of the electric magnet may be advantageously installed directly within the drum. The poles of the magnet or magnets extend in the peripheral direction of the drum. Hence, the windings have a longitudinal extending shape in the direction of the periphery.

It is, however, also advisable to provide several permanent magnets with high coercive power at a distance from one another and next to one another within the drum so that equal poles face the outer wall of the drum. In this way particularly great magnetic forces are produced. In this connection it is advisable for the permanent magnets to be attached direct to the inner wall of the drum and to rotate with the latter and further to provide a scraper in front of the suction intake of the conveyor tube.

In order to promote picking up of the magnetizable particles carried along by the outer surface of the rotating drum and attracted by the magnets, a scraper plate should be provided directly within the suction intake of the conveyor system. The suction intake of the conveyor system is so designed that a powerful current is produced within the range of the drum surface so that the particles are pulled along.

If several magnets or electric magnets, respectively, are provided next to one another in the axial direction it is advisable to arrange a separate suction intake above each individual magnet thus concentrating the main current share on those regions in which the magnetic attraction is greatest.

To avoid flushing away the picked up particles before they are seized by the suction intake, the suction intake should originate within the range in which the magnetic forces of attraction are still effective and it should continue to the range in which the magnetic forces of attraction have become small or disappeared. It is also possible to provide a crusher between the lifting system and the conveyor system for disintegrating large lumps of manganese.

Mineral Concentrator Device

Mineral aggregates, especially ores, that are found dispersed over a widespread remote area such as the ocean floor are gathered and at least partially separated

from silt to concentrate minerals for transport to the sea surface. *A.F. Sullivan;*
U.S. Patent 3,802,740; April 9, 1974; assigned to The International Nickel Co.
designs a mineral concentrator which is adapted for movement over the deep sea
floor and has means for gathering and separating dispersed mineral aggregates to
provide a concentration of desired minerals for transport to the sea surface.

The vehicle body has a flat bottom surface extending transversely over the horizontal width of a preselected path. The apparatus compacts mineral aggregates
and sea floor silt in front of the vehicle body down into the sea floor when the
vehicle is moved forward on the sea floor in the preselected path thereby forming a concentration of mineral aggregates in a subsurface silt zone.

The apparatus shears from the sea floor at least a portion of the mineral aggregate concentration when the aft portion of the vehicle moves over the concentration. A portion of the mineral aggregate concentration that is sheared loose
is then moved upward and guided to a preselected location where the aggregates
can be transported to the sea surface. Undesired silt is then separated from desired aggregates and silt is discharged from the apparatus while the apparatus is
on the ocean floor.

The apparatus also comprises a water guide and power driven impeller for directing a flow of substantially silt-free water through the separation and discharge
devices to wash silt forcibly from the upwardly moved portion and assist in discharging silt from the apparatus while on the sea floor.

Tractored Mining Plant Connected to a Sessile Ship

A deep sea mining system for mining ore from the surface of an ocean bed using
a mining plant that travels the bed, a sessile ship, and an interconnecting hoist
pipe circuit is designed by *E.B. Dane, Jr.; U.S. Patent 3,811,730; May 21, 1974.*
The system is transported to the mining site by a ship with streamline and structural strengthening sections which are later detached and reassembled into an
auxiliary vessel. The hoist pipe circuit is towed to the mining site in sections
which, during assembly of the hoist pipe circuit, are used to lower the plant to
the ocean bed.

After being lowered, the plant covers the bottom, the sessile ship following the
course of the plant, continually hovering over it. A scraper bucket system
mounted on arms incorporated by the plant scrapes ore nodules from the surface
of the bed. The plant cleans the nodules of valueless mud, crushes them to particle size and couples the ore particles to the hoist pipe circuit where they are
lifted as slurry and deposited in bins of the hovering sessile ship.

This system for mining at ocean depths of 100 fathoms or more comprises a
tractored central mining plant and a sessile ship interconnected by a string of
buoyant hoist-pipe sections. Except for the pipe sections which are carried in
tow, the system is transported to the mining site in the sessile ship. To reduce
drag and provide structural strengthening, the ship is equipped with streamlining,
strengthening and propulsion sections that are detachable at sea and may be reassembled into an auxiliary vessel. Upon reaching the mining site, the sections
are separated and the sessile ship is erected.

The mining plant is lowered by linking it up with the first and succeeding pipe sections one at a time. The hoist pipe circuit so formed is completed when the miner reaches the bottom and rests on the ocean floor.

The miner incorporates a tractored vehicle with a pair of arms projecting from its sides. Each arm carries a series of scraper buckets that ride on an endless chain that circulates along the length of the arm. As the vehicle travels the bottom linked with the hovering sessile ship by the hoist pipe circuit, the circulating buckets scrape nodules on either side of the vehicle cutting a wide swath in the deposit.

Gathered nodules are cleaned of valueless mud and crushed into particle size by apparatus on the vehicle. They enter the hoist pipe circuit as slurry and are pumped to temporary storage bins in the ship to await subsequent transfer to an ore freighter.

Because the system is capable of sweeping large areas, it recovers ore at a high rate. The system is efficient as the ore is cleaned of mud prior to transport thereby reducing power consumption. Operation is continuous and involves no turn-around time except for initial deployment and final recovery of the miner which may be some weeks apart. The system is self-contained, all necessary power, equipment, and facilities being supplied by the sessile ship.

The surface components of the mining system are provided by a tender that performs a number of functions to sustain the venture, including the following:

    transports the miner to a distant mining site;
    supplies power to the miner and hoist pipe circuit;
    provides a control facility for all apparatus in the system;
    stores ore mined from the ocean bed temporarily pending transfer
      to another ship;
    provides living accommodations for the crew and holds enough
      commissaries and fuel for at least a month's work; and
    provides a sufficiently stable and suitable base for lowering the
      miner and construction of the lengthy hoist pipe circuit there-
      with, for retrieving them upon conclusion of operations, and
      for receiving ore as it hovers over the miner working the ocean
      floor.

While conventional ships are adequate in some of the above respects, they may be troublesome with regard to the provision of a stable base in rough seas. During construction, the miner is lowered from the ship attached to the hoist pipe circuit as the latter is assembled from the floating pipe sections. The accumulating load may develop a mass of over 1,000 tons before the miner reaches the bottom.

A large component of the static load is compensated by buoyancy tanks built into the respective pipe sections. However, the dynamic load produced by the massive ship as it rolls, pitches and heaves in the waves, some of which are in the order of 20 feet or more in height, is imposed fully on the construction hoists and the uppermost sections of the pipe circuit. Such dynamic loads could damage the pipe circuit or cause it to break away from the ship altogether. The sensitivity of the hoist circuit to unwanted motion characteristic of conventional ships may be reduced by using complex compensating equipment in the pipe circuit. Alternatively, the surface terminal may be supplied by a semisubmerged catamaran, like those used on oil drilling sea rigs. However, the preferred form of this system features a simpler and less expensive approach by incorporating a surface ship whose ratio of displacement change to total displacement as a function of depth from an assigned waterline is small compared to that of a conventional ship.

Such features are realized in a sessile or flip type ship which has a relatively thin stem section and an assigned water-line midway up the stem. The section comprises a relatively narrow cylindrical steel shell bounded by two widening conical shells, with a slope of about 45°. The upper conical shell merges into a cylindrically shaped double decked top while the lower conical shell is bounded by a long and wide cylindrical shell that is submerged when the ship is erected along the vertical.

The cross-sectional area of the bottom cylinder is much greater than that of the stem section which is at the ocean surface. Consequently, change in displacement due to wave action is significantly less than that of a conventional ship. Moreover, the massive ship's submerged bottom provides substantial damping to motion, particularly heave, caused by waves.

The sessile ship has a draft of about 85 meters, which makes her too deep to float in many harbors. Therefore, she is conveyed from port to deeper water in a horizontal orientation. Because of the flatness of her top deck, she develops considerable drag or resistance to passage in the sea. Her velocity is accordingly limited to a few knots in this orientation.

With the aid of special power drives adapted to propel the ship at inclined attitudes, say with her top deck a few feet or so above the surface, the ship with miner attached may be transported to a mining site at restricted speeds. As surface conditions change from calm to rough, the inclination of the ship may be varied by internal ballasting so the top deck clears the waves. However, due to the slow speeds of such self-propelling modes, mining sites attempted are limited to short distances from port.

Since many of the productive mining areas may be some thousands of kilometers from port, it is desirable to provide for more expeditious transport to the distant sites. Accordingly, the sessile ship is supplied with special sections that reduce her resistance to motion, provide propulsion, and add structural strengthening to her weak stem section.

Harvesting Vehicle with Variable Bouyancy Sections

B.G. Stechler; U.S. Patent 3,812,922; May 28, 1974 describes deep ocean mining, mineral harvesting and salvage vehicles including a body integrally formed of a positive buoyancy material and having recesses therein to receive a plurality of variable buoyancy tanks. Eduction and coring mining systems are alternatively provided for the vehicles and the vehicles are propelled along the floor of the ocean by means of high velocity jets and/or turbine wheels. Briefly, this deep ocean mining, mineral recovery and salvage vehicle includes a body integrally formed with a predetermined configuration of a material having a high positive buoyancy and of substantial strength to withstand pressures to which the vehicle will be subjected at great ocean depths. The positive buoyancy material has an extremely low specific density in order to enhance ascent of the vehicle to the ocean surface upon completion of a submarine mining or salvage operation.

The configuration of the positive buoyancy body defines a plurality of cylindrical recesses opening at the top of the body to receive variable buoyancy tanks which are operative to control ascent and descent of the vehicle. The variable buoyancy tanks each include a pressure chamber filled with a high pressure fluid and a storage chamber filled with a fluid, usually a liquid such as seawater which is readily available. The chambers are arranged to permit expansion of the fluid in the pressure chamber to force the fluid from the storage chamber at a high pressure.

Located centrally of the vehicle is a payload receiving chamber which receives the mined material (e.g., mineral nodules) and operates to separate the minerals (e.g., nodules) from the brine and other sediment as they pass into the payload chamber through the mining operation. The mining operation is performed in one form through a number of skirts each of which in cooperation with the ocean floor forms a chamber.

Eduction mining is performed within the chamber by opening a valve in the variable buoyancy tanks and permitting the fluid under pressure, preferably compressed air, to force the fluid in the storage chambers, preferably seawater, out of the variable buoyancy tanks. The seawater is directed through an eduction mining unit including a venturi type section which operates to draw material (e.g., nodules, sediment, ores, etc.) within the chamber formed by the skirt and ocean floor, up through a pipe passing through the vehicle and into the payload receiving chamber.

A portion of the seawater in the eduction mining system may be used to exert considerable pressure on the sea floor and thus stir up the sediment to facilitate drawing the sediment through the venturi and into the payload chamber. As the gas in the variable buoyancy tanks expands forcing the seawater out of the tanks, the density of air being much less than the specific density of the seawater, at the end of the harvesting phase the buoyancy of the vehicle will change causing the vehicle to begin its ascent.

The change in buoyancy of the vehicle is sufficient to overcome the increased density caused by the payload within the payload receiving chamber so that the vehicle and payload are raised to the ocean surface automatically as the mining cycle is completed.

The vehicle described is generally characterized as having: a body formed of a positive buoyancy material having recesses formed therein; variable buoyancy tanks disposed in the recesses to provide a negative buoyancy relative to seawater during descent of the vehicle and a positive buoyancy relative to sea water during ascent of the vehicle; a payload receiver carried by the body; and mining means for delivering minerals from a surface on which the vehicle rests to the payload receiver. This design allows the salvage vehicle to withstand great depths while maintaining sufficient buoyancy to permit the collection of large payloads.

The variable buoyancy tanks of the vehicle may easily be removed and filling the tanks may be facilitated by the use of compressed air and seawater. The vehicle is movable along the floor of the ocean which increases the area able to be mined. No outer shell is required for the vehicle thereby reducing the weight of the vehicle and increasing operating submergence depths while facilitating ascent of the vehicle with a payload.

Suctioning Material into Submerged Container

J.B. Laarman; U.S. Patent 3,815,267; June 11, 1974; assigned to NV Industrieele Handelscombinatie Holland, Netherlands describes a method by which material is sucked up from the bottom of a body of water through a suction pipe into a wholly submerged container. The container is filled with water to submerge and trim the container, and this water is pumped out to create the suction that raises the material and to balance the added weight of the material so that the container remains at a constant depth.

Material is discharged in a stream of water upwardly from the bottom of the container, and water is simultaneously admitted to the ballast tanks to maintain the container submerged. The container is in the form of two conical frusta that open into each other and are traversed by a vertical shaft for the various conduits. The container for the material is centrally disposed and the ballast tanks are peripherally disposed.

This method makes it possible for the suction apparatus proper to remain in one and the same area for quite a time, because the container serves as a loading station with a buffer stock of the material sucked up from the bottom, from which container the material can be carried off in a hopper craft or the like. It is further possible to operate in deeper water than was previously the case because the head, that is the vertical distance the material must be sucked up, remains limited to the distance between the container and the bottom.

The apparatus comprises a container with a suction pipe attached to it which is divided into at least one ballast compartment by walls. The ballast compartment has supply and discharge openings for water and a hopper space for the sucked up material from the bed of a body of water, which is connected with the suction pipe and with the suction side of a pump and can be connected with a delivery chamber which is connected on one side with a discharge pipe and is connected on the other side with the pressure side of a pump.

Both the suction pipe and the suction side of the pump are advantageously connected with or in the vicinity of the upper portion of the hopper space. In this manner, sucking up takes place through the hopper space on filling the hopper space, so that the pump will only suck up water and will not suck up a mixture of water and material from the bottom.

The pump is preferably disposed in a shaft extending through the hopper space while a connection passing through the wall of the shaft serves to connect the hopper space with the suction side of the pump. This makes it possible to draw a high vacuum, relative to the water around the hopper craft, in the hopper space with the aid of the pump.

In this form, the lower ends of the shaft and of the container can form the boundary of the delivery space, and the discharge conduit connected with the space can extend through the shaft.

For the purpose of facilitating the discharge of the container, the dividing walls inside the container are so constructed that they have a downwardly tapered shape so that in essence the container is divided into a funnel-shaped hopper compartment and a ballast compartment extending about the same, the narrow end portion of the hopper compartment having valve members for connecting the hopper space and the delivery space.

A passage can connect the delivery space and the delivery side of the pump, in which passage openings are provided, which open into the delivery space underneath the valve members of the ballast compartment.

On discharging the container, water is pumped through the latter passage through the delivery space and the discharge conduit. This water will mix with the flow of material coming from the hopper space through the valve members and will carry the material along so that no material will flow through the pump even on discharging the container. It is to be understood that one and the same water pump can be used for loading and discharging the container.

The shaft in the container preferably extends to the upper part of the container over such a distance that the top of the shaft will remain above the water surface under all conditions.

In order to maintain the container at a constant depth and well trimmed, it is provided that on filling a hopper space ballast water will simultaneously be discharged, while on discharging the container, ballast water will be fed thereto. To this end, for example, depth gauges may be provided by which valves are operated in the supply and discharge conduits of the ballast compartments.

When the container is loaded, the weight of ballast water pumped out of the container will simultaneously indicate what weight of material is sucked up from the bottom into the container; consequently the quantity of material sucked up from the bottom can be determined in a very simple manner.

The container preferably has the shape of a float, which in essence is comprised by two frusta with their bases against one another, and a shaft projecting above them so that the container has little flow resistance and will be little affected by the swell.

PROPRIETARY SYSTEMS FOR RAISING ORE

Air Injection Method

M.W. Smith and C.S. Kluth; U.S. Patent 3,526,436; September 1, 1970; assigned to Westinghouse Electric Corporation describe an air lift for an underwater mining system in which aggregates collected by a bottom collector are propelled through a conduit to a surface vessel by means of an air injection system wherein a pipe supplying compressed air is situated within the conduit carrying the aggregates. The pipe is placed within the conduit to maximize clearance between the air pipe and the conduit. Connected to the bottom opening of the pipe is a member containing a plurality of small apertures behind which is a rotating turbine operable by air pressure. A tool is connected to the end of the compressed air pipe and is rotatable with the turbine to remedy any possible clogging of the conduit.

The system includes a first conduit for transporting the collected material. Extending between two vertically spaced points, this conduit includes, or is immersed in, a fluid. A second conduit is disposed within the first conduit and includes a discharge portion at the lower end while at the upper end means are provided to supply the second conduit with a fluid such as compressed gas for introduction of that fluid into the first conduit. The discharge end of the second conduit is provided with means to introduce small air bubbles into the first conduit to insure higher efficiency of the air lift system.

Hoist Piping for Constructing Fluid Circuit

E.B. Dane, Jr.; U.S. Patent 3,736,077; May 29, 1973 describes pipe sections adapted for construction into a lengthy fluid circuit for hoisting mineral slurry from a mining plant working a deep ocean bed to a surface ship. Each pipe section possesses termini for facilitating the initial underwater construction of the circuit and strength for supporting the mining plant as it is simultaneously lowered to the bed.

A pipe section contains a pump for lifting slurry through its length and a dump valve for dumping slurry in the event a malfunction occurs in the circuit. A series of buoyancy tanks separated by cushions disposed along the sides of a pipe section contribute counterbalancing forces that largely neutralize the load that would otherwise accumulate in the circuit. The cushions are adapted to prevent high direct contact pressures between opposing surfaces of adjacent tanks.

This hoist piping system is designed to be used in conjunction with a mining system which comprises a surface ship and a mining plant connected by this system of hoist pipe sections connected in series with the ship. The plant works the bottom, gathering ore and feeding ore slurry to the hoisting system that conveys it to ore bins in the surface ship.

The hoist pipes should be easily and inexpensively transportable to the mining site which may be a considerable distance from port. The seabed to be mined ordinarily rests at ocean depths presenting a hostile environment to both man and machine. Then, the hoist circuit must be such to permit deployment of the entire system at more favorable ocean depths.

Once constructed, the circuit should be capable of transmitting ore slurry processed by the plant without placing excessive loads on the circuit or ship. The circuit should have special provisions for treating malfunctions without requiring recall of the entire mining system, and be capable of withstanding the stresses of its environment and those created by the mining plant. The system operates by using buoyant sections of hoist pipe with termini facilitating the serial interconnection of pipes underwater into a circuit.

A pipe section contains a pump for lifting slurry through its length and an electrically actuated dump valve controlled from the surface ship for rapidly dumping slurry in the event of a failure in the circuit. A series of buoyancy tanks
separated by cushions are located on the sides of the pipe and contribute counterbalancing forces that keep the load of the circuit within tolerable levels. The
cushions are adapted to prevent high direct contact pressures between opposing
surfaces of adjacent tanks.

Cyclical Operation of Containers

In the apparatus designed by *K. Holzenberger and O. Schiele; U.S. Patent
3,753,303; August 21, 1973; assigned to Klein, Schanzlin & Becker AG, Germany*
for hydraulically raising ore, a plurality of containers in a riser conduit cooper-
ates cyclically with a main suction pump and a flushing pump.

Control valves are cyclically operated so that the main suction pump exhausts
clear flushing water from one container, whereby a pressure differential is produced for raising an ore-water mixture in the lower portion of the riser conduit
to the level of the container, while the ore-water mixture in another container
is flushed through the upper riser conduit portion to the surface level.

Between two and four containers can be used for the cyclical operation so that
the ore-water mixture flows continuously out of the upper riser conduit portion,
while any contact between the material and the main suction pump and the
flushing pump is avoided.

The apparatus uses a flushing pump and a main suction pump, which pump only
clear water, while the ore-water mixture is raised by a pressure differential produced by the main suction pump. The pumps never have contact with the
raised ore pieces, or with water from which the ore pieces have been filtered.
Nevertheless, a substantially continuous discharge of the ore-water mixture at
the surface level is obtained.

For this purpose, a distributing station including a plurality of containers, preferably between two and four containers, is provided in a riser conduit, and the
main suction pump and the flushing pump are cyclically used to produce the required pressure differential for raising the ore-water mixture, and for flushing
the containers.

One or several main suction pumps, preferably submersed in the water, and one
or several flushing pumps, which may be located above or below the water surface, are preferably provided and connected by suitable connecting conduits with
the containers at the distributing station.

One form comprises a riser conduit which includes a lower conduit portion having a lower end for picking up material from the bottom of the sea and an upper
conduit portion having an upper end located at the surface level of the sea; a
distributing station between the lower and upper conduit portions, and including
a plurality of containers, and a main pump; a flushing pump positioned for pumping clear water, preferably at the surface of the sea; a connecting conduit connecting the inlet of the main pump, the outlet of the flushing pump, and the

lower and upper conduit portions with the containers; control valves are located in the connecting conduit between the containers on the one hand, and the main pump, the flushing pump and the conduit portions on the other hand. The control valves are cyclically operated for effecting exhaustion of clear flushing water from one of the containers for producing a pressure differential between the lower end of the riser conduit and the container for raising a mixture of the material, such as ore pieces, and water, in the lower conduit portion of the riser conduit.

At the same time, the mixture is flushed from another container by the flushing pump into the upper conduit portion of the riser conduit and out of its upper end. As a result, the main pump and the flushing pump pump no solid material, but only clear water, while the mixture of solid material with water flows substantially continuously outward the upper end of the riser conduit means.

Preferably, the reduction of the pressure, the filling of the container with the mixture, and the flushing of the container, are cyclically carried out with a plurality of containers, so that one container is flushed while another container is exhausted to produce the pressure differential or filled with the material due to the pressure differential. In this manner, the flow of the flushed mixture in the upper conduit portion to the surface level is substantially continued.

An air venting conduit may be provided between the container and the air above the sea surface so that the pressure in the container is reduced to atmospheric pressure. However, it is also possible to exhaust clear flushing water from a container to create in the same a low pressure approaching the vapor pressure limit of water.

The arrangement of the process, in which no pump has contact with solid material raised from a great depth, permits the use of standard inexpensive pumps, which are not subjected to the wear caused by ore pieces, sand or abrasive substances floating in the water. The control valves in the connecting conduits are also protected from raised solid material.

Control of Turbulence and Velocity in Air Lift Hydraulic Pipe

In the process and apparatus designed by *J.G. Santangelo, M.A. Dubs and C.E. Schatz; U.S. Patent 3,765,727; October 16, 1973; assigned to Kennecott Copper Corporation* for transporting mined deposits from the sea floor, air is withdrawn from an air lift hydraulic pipe to control turbulence and velocity of flow through the pipe. The air is withdrawn at a location substantially above the air injection station where the expanding air causes excessively high velocity and turbulence which cause nodule breakage and wear of the pipe.

In one form a sleeve valve is used to selectively open perforations communicating with the interior of the pipe. The body of the valve can be connected to a vacuum source at the surface to increase the amount of air removed from the pipe. In a second form a perforated member is located within the pipe and flow of air to the outside of the pipe is controlled by valves. The valves also permit selectively connecting the perforated member to a vacuum source at the surface. Hydraulic control arrangements permit regulating the valves from the surface vessel to control the rate of removal of air or gas from the pipe. A well-known technique for transporting mined materials from the sea floor to the surface is the air (or gas) lift hydraulic dredge. In this air lift system, air or gas is injected into the dredge pipe or lift pipe at a location somewhat below the water surface and the injected air creates a lift effect which causes water and the mined particulate material to flow upwardly through the pipe.

The depth at which the air or gas is injected depends on the depth at which the mining is done so the necessary velocities are created within the pipe to transport the particulate material to the surface.

The air lift system presents several problems. First, in deep mining, i.e., at depths of 10,000, 15,000 ft or greater, it is necessary to inject the lifting air at some distance below the surface of the water, for example, 3,000 ft. A bubble of air at such depth will expand to almost 100 times its original volume during travel upwardly from the 3,000 ft depth to the water surface.

In addition, 50% of the total expansion of the bubble will take place in the final 34 ft of the ascent, where the static head and back pressure on the bubble is the lowest. Such tremendous expansion in the final section of the pipe leads to unfavorable conditions such as a very high exit velocity of the transported material and extreme turbulence within the pipe.

The high exit velocity from the pipe has required the use of baffles and other equipment to decrease the exit speed of the mined material. Such baffles, however, tend to cause the mined material such as manganese nodules to break and otherwise disintegrate, making them more difficult to handle and to separate undesirable materials, such as sand from the nodules. In addition, the turbulence and high velocities within the pipe cause further disintegration of the nodules.

This method of controlling the velocity and deleterious effects of the expanding air in the upper section of the lift pipe is accomplished by controlled removal of air from the pipe at one or more locations along the length of the pipe above the location where the lift air is initially injected or introduced. By bleeding air or gas out of the pipe at a controlled rate at one or more selected locations, the expansion of the air or gas within the pipe is controlled and correspondingly, the exit velocity of the mined material as well as the velocity of the material within the pipe is controlled.

Correspondingly, the desired lift is obtained to cause a flow of water through the pipe at a sufficient velocity to transport the mined materials from a location below the point of injection of the lifting gas, a three phase slurry of gas, water, and mined material exists in the pipe at a location above the injected gas, and the desired velocities within the pipe are maintained by selectively removing some of the gas or air from the pipe at a location below the uppermost portion of the pipe to control the expansion of the gas.

Also provided is an improved air lift hydraulic system for controlling the velocity of a slurry within an air lift pipe to reduce disintegration of particulate material transported by the pipe.

FINE SCALE SURVEY TECHNOLOGY

Over the last ten years a comprehensive capability has been developed for conduct of fine scale, precision, sea floor survey. The system includes a narrow beam echo sounder for topographic measurements; side-looking sonars having several scales of resolution to determine distribution of bottom roughness and location and shape of such features as rock outcrops or small scarps; 3.5 kHz echo sounder to delineate details of shallow structure; stereo photography to give a view of small-scale surface structure; and proton magnetometer to determine magnetic anomalies.

A transponder navigation system provides nagivation with residual uncertainty of a few meters. The system can be used from conventional research ships and has been operated to depths of 7,000 meters.

An unmanned, mobile device which operates on the sea floor to take samples and make in situ measurements of sea floor properties has also been developed. This latter equipment has not been as extensively used but provides a base of engineering and operational experience to combine with deep tow technology to produce a very powerful system for documenting the detailed nature of the deep sea floor.

PROPRIETARY EXPLORATION SYSTEMS

Explosive Vibration

W.P. Holloway; U.S. Patent 3,828,886; August 13, 1974 describes a geophysical exploration apparatus in which an explosion is set up to propagate vibrations which pass down into the earth. The explosion is controlled in such a manner that it blasts down into and through a body of water so that it does not disrupt the elasticity of a bored hole or segregated space in which the apparatus may be sealed off to contain the explosion.

The apparatus sets in motion vibrations or sound waves which pass down into the earth from a bore where a controlled explosion takes place in a manner that the waves are propagated in smooth patterns not calculated to disrupt the elasticity of the bored hole.

The vibrations, as reflected back to the earth's surface, are received by encased geophones with substantially wide bases of reception whereby the vibrations are unimpeded by extraneous noises which ordinarily occur in the weather zone just under or in the space just above the earth surface.

The apparatus comprises a detonator having an upper chamber into which a fuel and a combustion supporting gas together are admitted. An ignition device is provided to ignite the mixture in the upper chamber, the explosion or combustion products being urged through restrictions provided between the upper chamber and a lower chamber, which has at least its lower end under water at the time of ignition and combustion. As the products of combustion are first urged through the restriction or escape passage into the lower chamber to expand and pass down through the water in and/or below the lower chamber, a check valve provided to communicate with the interior of the lower chamber below the escape passage first remains closed to open as the pressure falls to permit air to be drawn into the lower chamber so that a vacuum does not result following the blast. Also provided is a device to admit a purge gas into the upper chamber down through the escape passage.

Earth Penetration in Deep Water at Atmospheric Pressure

P.W. Peter; U.S. Patent 3,830,068; August 20, 1974 describes a system for earth penetration in deep water at atmospheric pressure. The system enables underwater mining or other earth penetration to be conducted at atmospheric pressure in deep water through a hollow cylindrical entry column formed of concentric, radially spaced, cylindrical tubes extending from above water into the bottom and containing in the spaces therebetween water columns of progressively decreasing depth toward the center for offsetting the static pressure of the surrounding water.

Constructible in situ, the column serves as a monopod adapted to support above water a platform on which any suitable structure can be mounted. The column is stabilized against dynamic lateral forces by jets, anchors or other counteracting devices and is removable for salvage or reuse elsewhere on completion of a particular operation.

In the system a plurality of axially coextensive radially spaced concentric cylindrical thin walled tubes are made into an open ended cylindrical entry column having a hollow interior and a length to extend from above the surface into the bottom of a body of water at a location to be penetrated. The column is sunk vertically into the water with the tubes open at both ends for enabling the tubes during lowering to be continuously filled with water from below to the level of the surface.

The lower ends of the tubes, on reaching the bottom are thus able to penetrate into the bottom under the weight of the column and thereby close the lower ends. The column is thereafter partly pumped out to empty the interior. The water is pumped down between the tubes to form water columns of progressively decreasing height toward the center of the tubes for correspondingly reducing toward the center the hydrostatic pressure exerted on the exterior of the entry column by the surrounding water.

The hollow interior of the entry column forms a cylindrical access tube extending from the surface to the bottom of the body of water and includes a means for injection sealing in the bottom of the access tube to resist seepage of water from beneath the entry column.

In the system adjoining tubes are connected by a cross brace for radially spacing the tubes. Each tube is selected to have a bending resistance sufficient to withstand the hydrostatic pressure exerted on them by the difference in the heights of the water columns on their opposite sides. The bending resistance of the tubes is combined with the progressive decrease in the heads of the water columns therebetween to reduce the static pressure on the exterior of the entry column substantially to atmospheric pressure over the length of the hollow interior between the surface and bottom of the water body. The entry column is constructed on the surface and sunk progressively into the body of water. A platform is constructed above water on the upper end of the entry column, and a tank is installed on the platform for adding water loading to the downward force urging the lower end of the column into the earth.

MINERAL EXTRACTION

Implications of Internal Structure of Manganese Nodules

There are important advantages in studying polished cross sections of entire nodules. Whole nodule sections not only reveal small textural features but also the distribution and relative abundance of major mineralogical zones.

Many nodules are not homogeneous, and it may not be any wiser to process all nodules alike than it is to treat all ore veins in a mine in the same way. Processing must be carried out on thousands of tons of material, and not on individual nodules. If it can be shown that uniquely zoned nodules predominate in certain regions the following possibilities should be considered.

Optical study of nodules can be used to determine the form and extent of different mineralogical and chemical components of the nodules, and tests could be made to try to concentrate physically certain parts of nodules, for example, the material known to contain the highest percentage of Ni, Cu, and Mn. Of course, the same study may instead reveal that physical concentration is economically unfeasible.

If physical concentration processes are attempted, the microscopic study of mill products may be just as useful in judging results and resolving problems as it is in the treatment of many conventional types of ore.

If chemical separation of metals appears necessary, microtests of various kinds could be performed directly on the polished cross section of a typical nodule to find the most effective reagents, etc., for different kinds of nodule material. Microscopic examination of the results could lead to a rapid evaluation of different possible processes.

Routine microscopic study of representative polished sections may disclose the presence of certain minerals of unusual value which would be completely overlooked in the bulk crushing, grinding, and processing procedures now in use. For example, certain poorly crystalline but discrete minerals like nsutite, the best known natural battery depolarizer, may be found in nodules from some region; but chemical analyses would never show its presence, nor would large-scale x-ray diffraction analyses.

The structural region concept may be useful in judging the validity of interpolation between sampling localities far apart on the sea floor. For example, if adjacent sample sites are 50 km apart and bulk analyses show a marked similarity, it would be tempting to interpret the nodule deposits represented as continuous between the two sites. Statistically, however, the justification of such an interpretation is very tenuous. If, however, it were found that representative nodules from each site showed the same internal structure, mineralogy, and zonal succession, a similar growth history for nodules over the distance in question would be indicated; and interpolation between sampling sites would probably be justified (with due allowance for the effect of bottom topography).

In addition to these specific advantages for polished section and related study, it is not unreasonable to expect that this sort of routine investigation may from time to time reveal unexpected new facts about manganese nodules. The knowledge thus gained may lead not only to more efficient use of this great potential resource but also to new ideas for minimizing waste and protecting the natural environment of the sea and land.

Separation of Nickel from Cobalt Based on Phase Characteristics

J.L. Mero; U.S. Patent 3,169,856; February 16, 1965 describes a process for the separation of nickel from cobalt in ocean floor manganese deposits. These ocean floor mineral deposits are found to consist of several separate mineral phases of manganese and of iron. These separate mineral crystallites of manganese and of iron are so fine grained and so intimately mixed that no known physical process can be employed to separate them.

The minor metallic constituents of the ocean floor manganese deposits, such as nickel, cobalt, and copper and other elements, are localized in different mineral phases of this material, some elements being contained in the manganese mineral phases and others in the iron mineral phases.

The nickel and copper are contained in the manganese mineral phases of this material, apparently in solid solution in the manganese minerals. Cobalt, on the other hand, exists in the iron mineral phases of this material to the exclusion of the manganese phase. This is unlike the situation existing in connection with conventional ore involving nickel and cobalt, in which ores the nickel and cobalt are intimately mixed and must be indiscriminately leached from the ore, as the first step in effecting a separation of them, which places them both in solution together, which in turn raises the problem of separation of the nickel from the cobalt.

The process is predicated upon the phase characteristics of the ocean floor mineral deposits under consideration, and basically involves a process for differentially leaching these metals from the manganese ore before they are both in solution together.

To accomplish such separation of the nickel from the cobalt, the ore is first crushed preferably to about minus forty mesh. The crushed ore material is then mixed with water to form an aqueous slurry which consists of not more than about 40% of crushed ore by weight.

The amount of crushed ore in this slurry is not critical but should be sufficient to allow the mixture to act fluid. Leaching gases, which may be sulfur dioxide, nitrogen dioxide or other gases that reduce manganese and iron oxides of a high oxidation state, are then mixed with the aqueous slurry. This may be accomplished by percolating the gas up through the slurry while maintaining the slurry in a constant state of agitation. When the slurry is thus exposed to the gas, the manganese along with the nickel, copper and other mineral elements bound up in the manganese phase of the ore, will go into solution in accordance with the leaching curve of Figure 2.7, designated as "Leaching Curve for Ni-Cu in Mn Phase," while the iron along with the cobalt bound up in the iron phase, will leach out in accordance with the leaching curve identified as "Leaching Curve for Co in Fe Phase."

A comparison of the two curves establishes that the nickel, copper and such other elements in the manganese phase, will go into solution before any appreciable amounts of the cobalt and such other elements as may be bound up in the iron phase. After all the available manganese, nickel, copper, etc., are in solution, additional amounts of gas added to the aqueous slurry, will cause a change in the pH of the slurry toward the acidic and only then will much of the cobalt or iron start to dissolve.

It is at this point that the process is stopped and the separation effected, because at this point the nickel and associated elements are all in solution, while the cobalt and its associated elements are still in solid state. This critical point can thus be controlled by continuous monitoring of the pH of the aqueous slurry.

Another indication of the complete dissolution of the nickel and other metals associated in the manganese phase will be the evolution of leaching gas from the top of the leaching cell. Evolution of such leaching gas or the sudden change in the pH of the slurry, therefore, can be used as controls to prevent the dissolving of the cobalt and other elements associated with the iron phase in the ore material.

FIGURE 2.7: COMPARATIVE LEACHING RATES OF MANGANESE AND OF IRON MINERALS

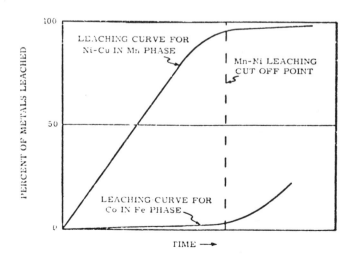

Source: J.L. Mero; U.S. Patent 3,169,856; February 16, 1965

This leaching process may be batchwise or continuous. At the completion of the leaching operation in a batch process, the leaching gases are stopped and the solution is separated from the tailings. In a continuous process, solution would be continuously added and the tailings continuously removed. The pH of the solution and the evolution of leaching gases would be closely monitored to determine the rate at which new slurry is added to the system and the rate at which solution is withdrawn.

The amount of leaching gas required to complete the dissolution of the elements in the manganese phase can be determined by calculation, but, in practice, a small excess of gas is always required to assure complete dissolution of the manganese, nickel, copper, etc., and the controls previously described, are of practical value in determining when these elements have been completely dissolved.

The rate at which the manganese, nickel, and associated elements are dissolved will depend somewhat on the temperature, pressure, and agitation of the slurry, but such variables are not critical nor essential features of the process.

There need be no external control of the pH of the aqueous slurry or resulting solutions. The pH of the slurry will be controlled by the reaction between the gases, the water, and the dissolving elements, and will be on the acidic side of 7, while the manganese, nickel and associated elements are dissolving. The value of the pH will depend somewhat on the rate of the reaction and the excess of the leaching gas used, but the exact value of the pH is not critical.

The resulting solution of manganese, nickel, copper, etc., is separated from the resulting cobalt-iron solid tails, before the iron and cobalt can dissolve appreciably. In a batch process, the cut off point of the leaching process would be at the point at which the cobalt or iron content of the resulting solution begins to rise rapidly, as illustrated in Figure 2.7.

The iron-cobalt tails are then removed from the vessel, dewatered, filtered, and washed, and then treated in a standard manner with acid to leach the cobalt from the iron. The leaching of the cobalt from the iron with acid is a standard process in the metallurgical industry.

The solution of manganese, nickel, copper, etc., can be treated with hydrogen sulfide to precipitate the copper, which is then separated from the remaining manganese-nickel solution by standard dewatering and filtering techniques. Nickel can then be separated from the manganese by precipitation with hydrogen gas, or the manganese can be first separated from the nickel by differential thermal reduction. The separation of the nickel from the manganese or the manganese from the nickel would be conducted according to standard procedures.

The makeup of the manganese ores found strewn over the ocean floor, lends itself to other methods of differentially leaching to effect separation of the nickel from the cobalt. One such additional method involves subjecting the ore material to preleaching reduction roasts. The common factor in carrying out this process, however, is to effect a differential leaching of the ore material, regardless of the particular steps carried out in effecting such differential leaching.

Mixed Nodule Ore and Iron Sulfide Mineral Treatment

W.S. Kane and P.H. Cardwell; U.S. Patent 3,809,624; May 7, 1974; assigned to Deepsea Ventures, Inc. provide a process for simultaneously obtaining the metal values from ocean floor nodule ore and from an iron sulfide mineral. The nodule ore comprises primarily oxides of iron and manganese plus nickel, copper and cobalt compounds.

The ore is reacted with the iron sulfide ore, which may also contain desired nonferrous metal values at elevated temperatures in the presence of oxygen to form iron oxide and the water-soluble sulfates of the nonferrous metals. The mixed reacted ores are then leached with water to obtain a solution and the resultant pregnant leach solution is treated, as by liquid ion exchange processes, to obtain separate streams of the individual metal values of nickel, cobalt, copper and manganese. The metals can be obtained by cathodic reduction of the compounds.

It has now been determined that by combining an ocean floor nodule ore with an ore containing a large proportion of iron sulfide, a most useful refining process for both ores is obtained, wherein the oxidizing property of the nodule ore and the reducing property of the iron sulfide ore are utilized.

The valuable nonferrous metal values from both ores are obtained as the water-soluble sulfates and the iron is in the form of insoluble iron oxide. This process provides a simple one-step method by which the iron can be separated from the nonferrous metals and the nonferrous metal values can then be readily separated into the individual metal values in order to obtain the desired elemental metals in a pure state.

This process provides means to refine oxidic ocean floor nodule ore, the ore comprising primary quantities of oxides of manganese and of iron and secondary quantities of compounds of nickel, cobalt and copper. The ocean nodule ore is admixed with an iron sulfide ore which comprises primary quantities of an iron sulfide compound plus secondary quantities of the sulfide of a nonferrous metal. Examples of nonferrous metals present in iron sulfide ores include copper, zinc, lead, nickel, cobalt, cadmium, molybdenum, tin, arsenic, antimony, and bismuth.

The two ores are treated together by a process comprising: admixing and roasting the two ores in the presence of oxygen at a temperature sufficient to convert any iron sulfide present in the sulfide ore to iron oxide and to convert the remaining sulfides and oxides to sulfates; leaching the roasted mixed ore with water so as to obtain an aqueous solution of the soluble sulfates of the metal values obtained from the sulfide ore, e.g., copper, lead or nickel and the soluble sulfates of manganese, nickel, cobalt and copper, the metal values obtained from the nodule ore; and separating the aqueous solution from any insoluble residue from the ores.

The aqueous solution of the mixed soluble sulfate salts can then be further refined to obtain the individual metal values which can then be reduced as desired to form the elemental metals. Preferably, the desired metal values are separated by individually extracting each metal from the solution in seriatum and then treating the individual metal values in a known manner to obtain the elemental metal.

In a preferred form, the aqueous solution is contacted with a liquid ion exchange medium capable of selectively extracting one of the metal values from the solution. The liquid ion exchange medium is separated from the aqueous solution raffinate and the metal value then stripped from the liquid ion exchange medium to form an aqueous solution of the separated metal value.

The individual metal value can then be reduced by known means, for example, electrolyzing the aqueous solution of the individual metal value, or evaporating the aqueous solution to obtain the solid salt and reducing the pure solid salt by fused salt electrolysis, or by cementing from aqueous solution, with another metal or with hydrogen gas.

Preferably, both the ocean floor nodule ore and the sulfide iron ore are comminuted in a crusher or grinder to a particle size of not greater than about 10 mesh on the U.S. sieve scale and preferably in the range of 25 to 100 mesh. The two ores can be crushed or ground together or mixed after they are comminuted.

As complete mixing as possible is desirable and for that reason grinding or crushing the two ores together does aid in forming a fairly uniform mixture of the two. Even better mixing is of course obtained when a fluidized reactor bed is used for the roasting. The air blow supporting the fluidation tends to form an extremely uniform mixture of the two ores.

The roasting is optimally carried out at 400° to 600°C. The pressure in the reactor does not affect the chemistry of the reaction. If a fluidized bed is used, the pressure requirement is that sufficient to maintain the bed fluid. In a nonfluid bed, substantially ambient or slightly above ambient pressure can be utilized in order to maintain the flow of air through the bed.

When utilizing air to maintain a fluidized bed, sufficient oxygen is provided to the ore particles to sustain the desired reaction. Generally, at least a stoichiometric amount of oxygen, preferably as air, should be provided, but a substantial excess of air is preferably provided. When operating in a fluidized bed using air as the fluidizing medium, a large excess of air is present to insure complete reaction of all the metal values to form the desired soluble sulfate salts.

The two ores can be generally mixed in substantially equivalent proportions depending upon the relative amounts of sulfur and metal values in each of two ores. The most efficient results, of course, would be obtained by utilizing a stoichiometric quantity of the two ores; however, as this is practically difficult to achieve, it is preferred that there be a slight excess of the iron sulfide ore in the reactor to insure complete reaction of the more valuable ocean floor nodule ore.

The reacted mixed ores can then be leached with water by conventional leaching procedures, utilizing conventional leaching equipment for liquid-solid contact. The leach water should have a pH of not greater than about 3 and preferably not less than about 2. The leaching with water can be carried out at substantially ambient temperatures. However, as the reactor ore is hot, the leaching is done at initially, at least higher temperatures below the boiling point of water. Leaching, generally, can be done optimally at 25° to 60°C. The amount of leach water required generally should be sufficient to dissolve all of the metal values present in both ores.

The contact can occur in a single stage, but, preferably, occurs in several stages, as in a mixer-settler system. The pregnant leach solution can be separated from the remaining solid residue and finally filtered to insure complete removal of all particulate solid matter. The solid residue from the leaching includes the iron oxide and gangue, or insoluble residue, from the ore.

The pregnant leach solution containing the dissolved soluble metals from the two ores can be separated from the insoluble solids by any conventional liquid-solid separation procedure, e.g., filtering, decanting, thickening, etc. The leaching and solid-liquid separation can be carried out batchwise or continuously, but preferably by continuous countercurrent flow.

It is important that all the iron be converted to the oxide and thus not dissolved by the pregnant leach solution. Iron tends to interfere with the preferred methods for separating and purifying the individual metal values remaining in the pregnant leach solution.

The pregnant leach solution obtained following the leaching contains generally up to about 110 grams per liter of manganese values, but optimally contains manganese in a concentration of 25 to 100 grams per liter. The concentration of the other metals in the leach solution are proportional to their concentration in the two ores relative to manganese. Generally, the manganese salt has the limiting solubility so that substantially all of the remaining metal values can be leached out with manganese. This is especially true with regard to nickel, cobalt and copper.

Because of the rather complex mixture of materials which are obtained from such ocean floor nodules especially when mixed with the metals from the sulfide ore, many of the standard hydrometallurigical methods for separating out metal values are not directly applicable because of the presence of various interfering ions. Preferably, the following procedures can be utilized for separating at least the pure cobalt, copper, nickel, and manganese metal values from the pregnant leach solution.

For separating each of the copper, cobalt, and nickel values from the leach solution a liquid extraction procedure is most preferred. The liquid-liquid extraction procedure requires the use of an extracting medium which is readily separable from water, which is selective for extracting one or more of the metal values from the aqueous leach solution and from which the metal value can be readily stripped.

The extracting medium should be immiscible with water to improve the economic efficiency of the process. If the extracting medium were not immiscible with water, a substantial loss of the extracting reagent would occur during each extraction, by virtue of at least a partial solubility in the water phase and a loss of the extracting agent in the aqueous raffinate.

Extracting agents which are especially suitable because they are highly specific to the metal values in the leach solutions which are obtained, e.g., from ocean floor nodule ores, include, for example, certain substituted 8-hydroxyquinolines, α-hydroxy oximes and naphthenic acids. Representative compounds useful for ion exchange are: 7-octylbenzyl-8-hydroxyquinoline and 7-dodecylbenzyl-8-hydroxyquinoline.

The second preferred type of metal extractants are the alpha-hydroxy oximes. The liquid ion exchange agents, which are used for the extraction of copper, cobalt and nickel values are generally chelates and thus remove only the metal values from the solution, leaving behind the anions.

The extracting agent can be a liquid which is itself water-immiscible but generally can be dissolved in a solvent which is substantially immiscible with water. The oximes and hydroxyquinolines are at least partially insoluble in water. It has been found to be preferable to use them in solution in a water-immiscible solvent to form a water-immiscible extraction medium to prevent loss of the extraction agent in the aqueous raffinate.

Solutions containing from 5 to 30% by weight of the extracting agent are economically useful as being sufficiently active to remove the desired metal values selectively from the aqueous solution and being sufficiently dilute in the extracting agent so that substantially no extracting agent is leached out and lost in the aqueous raffinate.

Useful solvents include generally any inert hydrocarbons which are solvents for the extracting agent, per se, and for the metal chelate, or, extracting agent-metal complex, and which do not react with any of the other materials present, under the conditions of the extraction process. Generally, liquid aliphatic, cycloaliphatic, aromatic, cycloaliphatic-aromatic, aliphatic-aromatic or chlorinated such hydrocarbons are preferably utilized as the solvent-diluent for the extracting agent.

Optimally, the diluent-solvent has a specific gravity in the range of 0.65 to 0.95 and a mid-boiling point in the range of $120°$ to $615°F$ (ASTM distillation). Examples of suitable solvents include benzene, toluene, xylene, aliphatic and aromatic petroleum fractions such as naphtha and derivatives thereof and mixtures of these.

Treatment of Ore with Acidic Reagent in Absence of Oxygen

In a process closely related to U.S. Patent 3,809,624 *W.S. Kane and P.H. Cardwell; U.S. Patent 3,810,827; May 14, 1974; assigned to Deepsea Ventures, Inc.* provide a process for obtaining the metal values from ocean floor nodule ore. The ore comprises primarily oxides of iron and manganese plus nickel, copper and cobalt compounds.

The ore is treated with an acidic reagent, e.g., SO_2, in the absence of oxygen, to form the water-soluble salt, e.g., the sulfate, of manganese, only. The ore is then leached with water to obtain a solution of the manganese salt, which may be further treated to obtain manganese metal. The ore residue can be further treated to extract the other metal values.

The ocean floor nodules have a rather unique physical structure and chemical composition, especially including the fact that the primary metal value present in the most valuable nodule ores is manganese. It is thus desirable to separate out the manganese metal value from the other desired metal values as early in the refining process as possible; this early removal results in the low cost manganese recovery, with a minimum of refining steps, while at the same time, by removing the manganese from the other metal values before they are being processed, removes a possibly interfering metal ion from the system.

The process comprises the following steps:

(1)   Reacting in the absence of oxygen nodule ore with an acidic reducing agent, which is capable of forming the corresponding water-soluble salt of divalent manganese from any oxide of tetravalent manganese present in the ore, but which will not react with the oxidic compounds of nickel, copper and cobalt, in the absence of oxygen;

(2)   Leaching the reacted ore with water to form an aqueous solution of the water-soluble manganese salt;

(3)   Separating the aqueous solution of manganese salt from the remaining solid residue of the ore;

(4)   contacting the solid residue of the ore with a salt-forming reagent so as to form the corresponding water-soluble salts of nickel, copper and cobalt; and

(5)   Releaching the rereacted ore to form an aqueous solution of the metal salts of copper, nickel and cobalt, and separating the aqueous solution from the solid ore residue and from the iron compound.

The aqueous solution which is thus formed, comprising the metal values of the ore, except for the major part of the manganese and substantially all of the iron, can be further treated to separate the individual metal values of nickel, copper and cobalt into separate streams. The separated metal values can then be reduced to the elemental metals, as by electrolysis.

Steps (4) and (5) can be carried out concurrently when the salt-forming reagent is present in solution or the reaction is carried out in an aqueous medium. The events in Step (5) above can be carried out simultaneously or in chronological sequence. Encompassed within this step are processes wherein a solution of all the metal values, including those of iron, cobalt, nickel and copper are dissolved in water and the iron is then removed from the solution.

Also encompassed within Step (5) are processes wherein the iron is first converted to a water-insoluble material, e.g., iron oxide, before leaching with water, the remaining metals dissolved in water and the solution separated from the insoluble iron material.

If an aqueous solution is formed which contains any dissolved iron, the iron can be removed by a variety of means including: increasing the pH of the solution above 3 and passing oxygen therethrough to precipitate the iron as iron oxide; extracting the iron solute, as by liquid extraction; or drying the solution and then converting the iron salt to iron oxide at elevated temperatures above 200°C in the presence of water.

The iron-free aqueous solution of the metal values can be separated into the individual metal values, for example, by liquid ion exchange. The first stage of this process is a first order reaction and thus proceeds at a relatively fast rate regardless of temperature, e.g., at ambient or higher. The procedure can be carried out at 15° to 45°C. The reacted ore can then be leached with water, after it has been removed from the reaction medium; the water preferably has a pH of not greater than about 2.

The leaching with the water can be carried out also at substantially ambient temperatures with the preferred range being 10° to 45°C. The reacted ore, following leaching, is substantially free of tetravalent manganese. There may be a relatively small portion of divalent manganese present in the ore, plus the other metal values, e.g., nickel, cobalt and copper. The ore can then be treated in a variety of ways so as to separate the remaining valuable metal values from the ore.

The ore can be reacted with substantially any acidic medium, i.e., any medium which would react with the insoluble oxidic metal compounds in the ore to form product compounds of the metal values which are readily separated from the gangue, or detritus, of the ore. Such product compounds are water-soluble, and thus can be separated from the residual ore material by aqueous leaching.

For example, the leached ore can be next treated with a source of halide ion, such as a hydrogen halide, to form the water-soluble halides of cobalt, copper and nickel. Such halides can be separated from the reaction mixture by leaching with an aqueous liquid. The reaction can be carried out using vaporous hydrogen halide or an aqueous solution of hydrogen halide at a wide range of temperatures.

Another procedure for reacting the ocean floor nodule ores results in the formation of the sulfates of the metal values. For example, by reacting the leached ore with sulfur dioxide, in the presence of oxygen, the sulfates of the nickel, copper and cobalt values are formed. The reaction with the sulfur dioxide can be carried out in an aqueous medium, in which case the water-soluble salts of nickel, copper and cobalt are immediately formed and dissolve in the aqueous reaction medium, whereas the presence of excess oxygen prevents the formation of any soluble iron salt and thus the dissolution of any substantial amount of the iron into the aqueous solution. The aqueous solution can be separated from the solid ore residue, including the iron.

As another alternative, the leached ore can be contacted with an ammoniated solution comprising dissolved ammonia and a strongly negative, stable, soluble anion. The solution should be sufficiently concentrated in ammonia to permit the formation of metal-ammonia complexes of the nickel, cobalt and copper, but also sufficiently concentrated in the anion to prevent the formation of a complex including the hydroxyl anion.

The releach solution contains the metal values of nickel, cobalt, and copper from the nodule ore plus trace quantities of the other metal values which are present in the ore. There may sometimes be some manganese removed at this stage. The releaching of the nodule ore when utilizing sulfur dioxide as the salt-forming reagent can be carried out at substantially ambient temperatures. The preferred range of temperature is 20° to 50°C. Generally, sufficient leach and releach solution should be fed to the ore so as to form a solution which is less than saturated in the desired metal values.

The nodule ore is preferably comminuted, as by grinding, crushing or pulverizing into small particles of about 35 to 100 mesh. This tends to increase the surface area for reaction and, therefore, is desirable in improving the rate of the reactions. When carrying out the reaction in an aqueous medium, the comminuted ore is preferably slurried in the water and the slurry is then contacted with the reagent, for example, if a gas such as sulfur dioxide is used, the gas is bubbled through the slurry.

The first leach solution is a substantially pure aqueous solution of a manganese salt and thus can be directly treated to obtain elemental manganese metal. The manganese metal can be obtained by electrolysis of the aqueous solution directly. The second or releach solution, however, is a solution of a mixture of at least the three secondary metal values of the ore, i.e., nickel, copper and cobalt, plus traces of the other metal values present in the ore. This is a rather complex aqueous solution; as a result many of the standard hydrometallurgical methods for separating metal values are not directly applicable because of the various interfering ions.

Preferably, the following procedures can be utilized for separating at least pure cobalt, copper and nickel from the pregnant releach solution. For separating each of the copper, cobalt, and nickel values from the releach solution a liquid extraction procedure is most preferred. The liquid-liquid extraction procedure requires the use of an extracting medium which is readily separable from water, which is selective for extracting one or more of the metal values from the aqueous leach solution and from which the metal value can be readily stripped.

The extracting medium should be immiscible with water to improve the economic efficiency of the process. If the extracting medium were not immiscible with water, a substantial loss of the extracting reagent would occur during each extraction, by virtue of at least a partial solubility in the water phase and a loss of the extracting agent in the aqueous raffinate.

Extracting agents which are especially suitable because they are highly specific to the metal values in the leach solutions which are obtained, e.g., from ocean floor nodule ores, include, for example, certain substituted 8-hydroxyquinolines, α-hydroxy oximes and naphthenic acids.

Representative compounds useful for ion exchange are: 7-octylbenzyl-8-hydroxyquinoline, 7-dodecylbenzyl-8-hydroxyquinoline, 7-nonylbenzyl-8-hydroxyquinoline, 7-di-tert-butylbenzyl-8-hydroxyquinoline, 7-hexadecenyl-8-hydroxyquinoline, 7-dibenzyl-8-hydroxyquinoline, 7-dimethyldicyclopentadienyl-8-hydroxyquinoline, 7-phenyldodecenyl-8-hydroxyquinoline, and the like.

The second preferred type of metal extractants are the α-hydroxy oximes, which are disclosed in U.S. Patents 3,224,873; 3,276,863; and 3,479,378. The extracting agent can be a liquid which is itself water-immiscible but generally can be dissolved in a solvent which is substantially immiscible with water. The oximes and hydroxyquinolines are at least partially insoluble in water. It has been found to be preferable to use them in solution in a water-immiscible solvent to form a water-immiscible extraction medium to prevent loss of the extraction agent in the aqueous raffinate.

Useful solvents include generally any inert hydrocarbons which are solvents for the extracting agent, per se, and for the metal chelate, or extracting agent-metal complex, and which do not react with any of the other materials present, under the conditions of the extraction process. Generally, liquid aliphatic, cycloaliphatic, aromatic, cycloaliphatic-aromatic, aliphatic-aromatic or chlorinated such hydrocarbons are preferably utilized as the solvent diluent for the extracting agent. Optimally, the diluent solvent has a specific gravity of about 0.65 to 0.95 and a mid-boiling point of about 120° to 615°F (ASTM distillation).

However, substantially any liquid can be used as a solvent-diluent that meets the following criteria: a solvent for the extracting agent; a solvent for the extracting agent-metal complex, or chelate; immiscible with water; and readily separable from water. In addition to the diluent and the extracting agent, there can preferably also be present in the liquid extracting medium a phase modifier which prevents formation of an emulsion with, or entrainment of, the organic phase in the aqueous phase.

The following example shows a preferred form of the process. Ocean floor nodule ore having the following composition was obtained.

Component	Percent by Weight
Manganese	17.65
Iron	10.6
Nickel	0.65
Cobalt	0.32
Copper	0.12
Other metals	Traces

The nodule ore is ground to an average particle size of less than 100 mesh. The comminuted ore is passed into a reaction vessel where it is maintained in a fluidized condition by an upwardly flowing stream of gaseous SO_2. The reactor is initially at a temperature of about 25°C and at ambient pressure.

At startup, the vessel was first bled free of all air, and other oxygen-containing gases, utilizing nitrogen gas, prior to addition of the first SO_2 gas. The positive pressure in the reactor vessel also serves to prevent any leakage of air into the first system. The flow of SO_2 is substantially pure SO_2 gas.

The temperature of the bed increases during the process to about 100°C. The reacted nodule ore is then passed to a leach tank where it is contacted with water having a pH not greater than 5 in three mixer-settler stages; the overflow liquid from the final settling stage is then passed to a drum filter where any remaining solid residue is removed. The filtrate, which is substantially an aqueous solution of pure manganese sulfate ($MnSO_4$), is passed to an electrolytic cell to reduce the manganese sulfate to elemental manganese metal.

The solid residues from the final settling tank and from the drum filter are combined and mixed with approximately twice its weight of water into which is passed a mixture of air and sulfur dioxide. The total composition of the gas is approximately 10% SO_2, 15% oxygen. The mixture of reacted ore and water is then permitted to settle and the aqueous solution is decanted off and passed to a drum filter to remove any remaining residue. The liquid filtrate comprises an aqueous solution of nickel sulfate, cobalt sulfate, copper sulfate and manganese sulfate.

It was found that the proportion of nickel, cobalt, copper and manganese in this solution, combined with the manganese removed in the first stage of the process, was substantially 98% of these metal values in the original ore. There was substantially no iron found in the aqueous solution. The solid residue separated from the aqueous releach solution following treatment with SO_2 and oxygen contained the solid ore residue, or gangue, which is primarily calcium compounds, and the iron, generally as iron oxide. The aqueous releach solution of the metal sulfates was then treated with a liquid ion exchange medium to remove the individual metal values of copper, nickel and cobalt from the remaining manganese as follows. The releach solution contains manganese sulfate, copper sulfate, nickel sulfate and cobalt sulfate in solution.

This material was extracted utilizing a solution comprising 10% by volume of an α-hydroxyoxime (5,8-diethyl-7-hydroxydodecane-6-oxime, known as LIX-64N), 20% by volume isodecanol, and the balance a mixed hydrocarbon solvent, comprising mixed aromatic-aliphatic petroleum hydrocarbons having a boiling point range of 410° to 460°F, and a specific gravity of 0.81. The aqueous raffinate had its pH adjusted to about 2.

The pH was maintained at about 2 by the addition of caustic during the extraction of copper. The aqueous raffinate and organic extractant passed countercurrently through five mixer-settler stages at an organic-to-aqueous ratio of 5:1 by volume. The aqueous raffinate from the copper extraction contained substantially all of the manganese, nickel, and cobalt originally present in the releach solution, but substantially all of the copper had been extracted.

Following the separation from the final settling stage, the organic extract was stripped of copper by spent acid solution from a copper aqueous electrolysis cell having a hydrogen ion concentration of 3 N, in H_2SO_4 utilizing countercurrent flow through five stages of a mixer-settler series.

The aqueous raffinate from the copper extraction step was adjusted to a pH of about 4.5 by the addition of 2 N caustic solution. The resulting aqueous solution was extracted in a five stage mixer-settler system, with a solution of 10% by volume 7-[3-(5,5,7,7-tetramethyl-1-octenyl)]-8-hydroxyquinoline plus 20% by volume isodecanol in kerosene to extract nickel and cobalt.

The nickel was stripped from the organic extract phase using the spent solution from a nickel electrolysis cell to which sulfuric acid was added to a concentration of hydrogen ion of 3 N in order to insure stripping of all of the nickel. The organic liquid and stripping acid were passed countercurrently through three mixer-settler stages at an organic-to-aqueous liquid ratio of 3:1, by volume. Substantially all of the nickel was removed from the organic phase.

The cobalt was next stripped from organic extract phase utilizing an aqueous solution containing 20% by weight HCl, in four mixer-settler stages at an organic aqueous ratio of 3:1. The cobalt was extracted from the 20% HCl solution using a kerosene solution containing 10% by volume triisooctylamine (TIOA), in three mixer-settler stages at an organic:aqueous volume ratio of 2:1. The cobalt was stripped from the TIOA solution utilizing spent aqueous electrolyte from a cobalt electrolysis cell in three mixer-settler stages with a 1:2 organic:aqueous phase ratio.

There were thus obtained, as a result of this process, four separate final streams each containing substantially pure metal salt: copper sulfate, nickel sulfate, cobalt chloride and manganese sulfate. Each of these aqueous solutions could be further treated by known methods to reduce the salts to the respective elemental metal.

Manganese sulfate is preferentially reduced in an aqueous electrolytic cell. The copper, nickel and cobalt salts are preferably electrolyzed in aqueous electrolytic cells.

Fused Salt Electrolysis to Obtain Manganese

B.E. Barton and P.H. Cardwell; U.S. Patent 3,832,295; August 27, 1974; assigned to Deepsea Ventures, Inc. found that is is possible to obtain substantially pure molten manganese in a cell utilizing inert anodes and cathodes in which the molten electrolyte is comprised only of halides. The product from this cell is a pure manganese metal in which the presence of oxides is substantially eliminated and from which substantially all of the reaction products formed during the electrolysis are removed by vaporization, resulting in the production of manganese metal which is substantially pure.

The process comprises first forming a molten mixture of halides, comprising a manganese halide, a bath halide selected from the group consisting of alkali metal halides and alkaline earth metals halides, and a halide of a reactant metal. The reactant metal can be any cathodically reducible metal which can replace manganese from its halide, but is preferably magnesium or aluminum.

The mixed fused halide bath is electrolyzed to form manganese metal which sinks through the bath and collects as a molten layer at the bottom of the bath while halogen is collected at the anode. Heat is applied to the bottom of the bath to maintain the manganese metal in a molten state while the top of the bath is preferably maintained at a sufficiently low temperature to prevent the vaporization and loss of metal halides, but at least about 650°C. A temperature gradient can be maintained through the molten bath to avoid vaporization of any halides.

The apparatus in which the electrolysis procedure is carried out is a furnace wherein the upper portion of the furnace is a fused salt electrolytic cell and the lower portion is a molten metal collection section. The temperature maintained in the lower metal collection portion of the furnace is sufficient to maintain the manganese metal in the molten condition, but the temperature in the upper cell portion of the furnace is sufficient to maintain the metal halides in a molten condition, but is below the boiling point of the metal halide mixture. The electrodes are located in the upper, or cell portion, of the furnace and the flow of electricity is in a horizontal direction across the bath.

The potential drop between the electrodes is not sufficient to cathodically reduce the manganese halide to manganese metal. It is believed that the potential drop is sufficient, under the conditions of the process, to cathodically reduce the halide of the reactant metal to the elemental reactant metal. Preferably, the temperature of the halide bath is above the melting point of the reactant metal, e.g., 650°C for magnesium and 660°C for aluminum.

Preferably, the reactant metal is formed as small particles or droplets dispersed in the molten mixed metal halide bath. This provides a greater surface area for reaction between the reactant metal and the manganese halide. To further prevent the agglomeration of the molten reactant metal, an antiagglomerating agent is added, which serves to maintain the molten reactant metal in a dispersed state in the molten halide bath. This is especially useful when magnesium halide is present.

The bath halide and the halide of the reactant metal are miscible with manganese halide but immiscible with manganese metal. The bath halides are generally inert to the manganese halide under the conditions of this process and do not react with the cathodically reduced reactant metal, e.g., magnesium or aluminum. They are essentially inert with regard to manganese metal and generally, are noncorrosive to the usual types of high temperature ceramic liners available.

The bath halides must be a mixture of at least one alkali metal halide and at least one alkaline earth metal halide. The mixture can form a low melting eutectic composition so as to reduce the temperature, at least in the upper electrolysis cell portion of the apparatus.

Certain of the halides used, especially manganese halides and aluminum halides (if used as reactant halides) have relatively high vapor pressures under the conditions of the electrolysis cell. Aluminum halides, in the pure state, boil at temperatures substantially lower than those under which these cells are normally operated, e.g., about 650°C and higher.

However, when the aluminum halides are admixed with the alkali and/or alkaline earth metal halides of the bath halide, the vapor pressure of the mixture is substantially reduced, apparently by the formation of certain complexes, and the boiling point of the mixture at atmospheric pressure is raised to substantially above 650°C.

Preferably, there is present at least one alkali metal halide. Useful alkali metal halides comprise the chlorides, iodides, and bromides of sodium, potassium, cesium, rubidium and lithium. Useful alkaline earth metal halides comprise the chlorides, bromides, and iodides of calcium, barium, strontium, and magnesium. In determining composition of the molten halide mixture, the bath mixture, refers to the combined reactant halide and bath halide, i.e., everything except the manganese halides.

The bath mixture comprises from 30 to 60% by weight of the bath mixture of an alkali metal halide, about 10 to 25% by weight of a halide of a reactant metal and 15 to 60% by weight of an alkaline earth metal halide. It is understood by the above stated ranges of composition, that when, e.g., a magnesium halide is utilized as the halide of a reactant metal, the total proportion of magnesium halide present in the bath mixture can be from 10 to 70% by weight of the bath mixture.

In addition, there is preferably also present in the bath mixture a small proportion of an antiagglomerating agent, such as an alkali metal tetraborate, preferably from 0.001 to 0.01% by weight. There can be present 0.2 to 90% by weight of the total mixture of a manganese halide. Preferably, however, the manganese halide is present in an amount from 0.5 to 10% by weight.

The temperature of the salt bath should be just sufficient to maintain the halide and preferably also the reactant metal in the molten condition. It is desirable to avoid loss by vaporization of any of the volatile halides. This is especially true of the aluminum halide if used. Accordingly, the temperature of the top surface of the molten mixed halide bath can vary from 650° up to 1260°C. When aluminum halide is used as the reactant halide, at higher temperatures, the volatile halide can be lost by vaporization.

Accordingly, the preferred temperature for the top surface of the mixed halide
bath containing aluminum halide is from 800° to 1000°C. The molten manganese
layer at the bottom of the bath is preferably maintained in the range of 1260°
to 1300°C. Useful antiagglomerating agents include boric oxide or metal salts
of boric acids, for example, B_2O_3, $Na_2B_4O_7$ and $K_2B_2O_4$.

The halogen is removed at the anode. Under the conditions of the cell, all of
the halogens, i.e., chlorine, iodine or bromine are in the vaporous state and can
be readily removed. The reactant metal, e.g., aluminum or magnesium which is
formed remains in the salt bath and reacts with the manganese halide, displacing
the manganese to reform magnesium halide or aluminum halide and the elemental
manganese metal.

The elemental manganese metal sinks through the molten halide bath to collect
as a lower layer in the narrow neck, or furnace, portion of the cell-furnace combination. Thus, there is substantially no net loss of the reactant halide, e.g.,
magnesium halide or aluminum halide, from the cell.

The manganese halide must be replaced in the bath during electrolysis. The rate
at which the manganese halide is consumed is dependent upon the rate at which
the reactant metal is formed, which in turn is dependent upon the current flow.
The rate of feed of the manganese halide must be sufficient to maintain a sufficient concentration of manganese halide in the electrolysis bath to react with the
reactant metal to form manganese metal and the halide of the reactant metal.

Recovery of High Purity Molten Manganese

*W.S. Kane and P.H. Cardwell; U.S. Patent 3,832,165; August 27, 1974; assigned
to Deepsea Ventures, Inc.* provide a process for obtaining high purity manganese,
in the molten state, from manganese oxide ores containing iron. The process
comprises halidating the ore with a hydrogen halide and leaching to obtain a
leach solution comprising manganous and ferric halides and the elemental halogen;
extracting the ferric halide from the solution; separating anhydrous manganese
halide from the solution; reducing the anhydrous manganese halide in the molten
state with aluminum to form molten manganese metal and aluminum halide; reacting the aluminum halide with water vapor to form hydrogen halide and aluminum oxide; and recycling the hydrogen halide.

Usually, where the ore also contains a nonferrous metal more noble than manganese, the halide of such nonferrous metal is removed from the leach solution,
after the ferric halide is removed, by precipitation. Figure 2.8 is a schematic
flow diagram of the process utilizing aqueous hydrogen halide as the halidation
reagent.

In the process, a manganiferous ore is crushed to a particle size of not greater
than about 35 mesh U.S. sieve scale. The crushed ore then passes to the five
stage halidation reactor where it is contacted countercurrently with an aqueous
hydrogen halide solution, e.g., initial concentration 11 N HCl. The halogen, e.g.,
chlorine, by-product is vented from each stage. Additional hydrogen halide gas
can be added to one or more of the stages, if desired. The aqueous leach solution leaving the final stage has a pH of from 1 to 2 and contains dissolved, the
soluble metal chlorides extracted from the ore.

FIGURE 2.8: FLOW DIAGRAM FOR MANGANESE RECOVERY

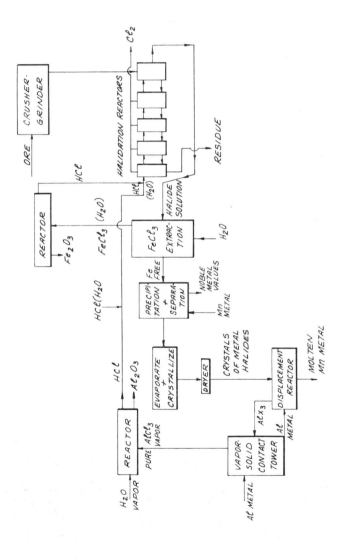

Source: W.S. Kane and P.H. Cardwell; U.S. Patent 3,832,165; August 27, 1974

The leach solution is next treated to an $FeCl_3$-extraction process which comprises being passed through 4 mixer-settler stages countercurrently to an organic solution of an amine extracting medium, e.g., comprising 15% by volume of N-lauryl-N-(1,1-dimethyleicosyl)amine,

$$[n\text{-}C_{12}H_{25}HN-\underset{\underset{CH_3}{|}}{\overset{\overset{CH_3}{|}}{C}}-(CH_2)_{18}-CH_3],$$

15% by volume isodecanol, in kerosene solvent, at an aqueous-to-organic ratio of, e.g., 1:4 by volume. The organic extract is then stripped with water, having a pH of 2, in a countercurrent, e.g., three stage mixer-settler system at an organic-to-aqueous ratio of, e.g., 4:1 by volume. The stripping solution is then passed to the $FeCl_3$ reactor, where the solution is evaporated, heated to a temperature of about 200°C to form HCl and Fe_2O_3. The HCl is passed to the halidation reactor.

The iron-free raffinate from the ferric halide extracting step is then passed over a bed of particulate manganese metal having a particle size not greater than about ¼ inch. The noble metal values are precipitated as the elemental metal on the manganese bed and manganese metal substitutes for the more noble metals in the halides, forming a solution more concentrated in manganese halide.

Any of the halides of the less noble metals present, such as the alkali metal halides or the alkaline earth metal halides, are not affected by this displacement treatment. The noble metal values precipitated in the bed of manganese metal can then be dissolved, as by an ammonium carbamate solution and then further treated to separate and purify the individual valuable metals obtained thereby.

The aqueous effluent from the metal precipitation step comprises manganese halide and usually also one or more alkali metal halides and alkaline earth metal halides. The effluent is next heated to crystallize the manganese halide tetrahydrate together with some of the other halides present.

The crystals are dried and added to a reaction vessel having a refractory lining, comprising a pool of molten halides (comprising about 50% manganese halide), wherein the top of the pool has a temperature of 1150°C and the bottom, 1300°C. A substantially stoichiometric amount of scrap aluminum turnings is added with the additional manganese halide. Molten manganese metal is tapped from the bottom layer of metal, through a liquid outlet port, and aluminum halide vapor removed overhead through a vapor outlet.

The aluminum halide vapor is passed through a contact tower with the aluminum scrap to preheat the aluminum and to recover any manganese halide carried out. The manganese halide is returned to the reactor. The temperature in the tower is maintained at above the boiling point of the aluminum halide.

The aluminum halide vapor is then passed into a reactor in contact with water vapor, at a temperature of at least about 400°C. Aluminum oxide is removed as a by-product and the hydrogen halide is recycled to the halidation reactor.

MANGANESE NODULES AS CATALYSTS

Use in Demetalation Process and Subsequent Removal of Metallic Constituents

P.B. Weisz and A.J. Silvestri; U.S. Patent 3,716,479; February 13, 1973; assigned to Mobil Oil Corporation describes the demetalation of a hydrocarbon charge stock which which involves contacting the hydrocarbon charge stock with hydrogen in the presence of manganese nodules as a catalyst.

The manganese nodule may be employed without pretreatment or may be pretreated by sulfiding or by leaching to remove and recover one or more valuable metallic constituents. The manganese nodule catalyst, after it has become deactivated by use, may be processed to remove and recover one or more valuable metallic constituents.

The chemical and physical properties of manganese nodules, as catalytic agents for the demetalation of hydrocarbon charge stocks, are, as compared with conventional catalytic agents for this purpose, considered to be somewhat unusual. The nodules have a high surface area, about 100 to 250 square meters per gram. They will, however, lose surface area by metal deposition during the demetalation reaction.

Further, as shown by R.G. Burns and D.W. Fuerstenau in *American Mineralogist,* volume 51, pages 895 to 902, "Electron-Probe Determination of Inter-Element Relationships in Manganese Nodules," the concentrations of the various metals contained in the nodules, i.e., the manganese, iron, cobalt, copper and nickel, are not uniform throughout the crystalline structure of the nodule.

Rather, a traverse across a section of a nodule will show marked differences in the concentrations of the various metals from point to point of the traverse. However, there appears to be a correlation between the concentrations of iron and cobalt. On the other hand, manufactured catalysts for demetalation are usually as uniform as the manufacturer can achieve.

The manganese nodules can be employed as the catalyst for the demetalation of the hydrocarbon charge stock substantially as mined, or recovered, from the floor of the body of water in which they occurred. Thus, the nodules, as mined, possibly after washing to remove seawater or lake water therefrom and mud or other loose material from the surface of the nodules, may be employed for demetalation.

The demetalation reaction may also be carried out employing, as the catalyst, manganese nodules which have been subjected to a pretreatment. Pretreatments to which the manganese nodules may be subjected include sulfiding or leaching to remove one or more components from the nodules.

Sulfiding of the manganese nodules increases the extent of demetalizing of the charge stock. It also can increase the extent of desulfurization and Conradson Carbon Residue (CCR) reduction, each of which is desirable. This treatment is carried out by contacting the nodules with hydrogen sulfide. The hydrogen sulfide may be pure or may be mixed with other gases. However, the hydrogen sulfide should be substantially free of hydrogen. The temperature of sulfiding may be from 300° to 450°F and the time of sulfiding may be about 4 to 8 hours.

The sulfiding may be effected, for example, by passing the hydrogen sulfide over the manganese nodules continuously during the sulfiding reaction. The space velocity of the hydrogen sulfide is not critical and any space velocity compatible with the equipment and such that some hydrogen sulfide is continuously detected in the exit stream is suitable.

The manganese nodules may also be pretreated by being subjected to leaching to remove one or more components. The nodules which contain, in addition to manganese, copper, nickel and molybdenum may be pretreated to leach the copper, nickel, or molybdenum, or any two, or all three, of these metals. The manganese nodules contain the copper, nickel, and molybdenum in sufficient quantities to provide a commercial source of these metals.

Further, the removal, at least partially, of these metals and other of the metallic constituents of the nodules has apparently no effect on the catalytic activity of the nodules for demetalation of hydrocarbon charge stocks. Thus, copper, nickel, and molybdenum, and other metals, may be recovered from the nodules for the economic advantage to be gained by such recovery and the remainder of the manganese nodules can then be employed as a catalyst for demetalation of hydrocarbon charge stocks.

Removal of the copper and the nickel may be effected by leaching the manganese nodules with an aqueous solution of a strong acid, such as hydrochloric, sulfuric, and nitric. The molybdenum may be removed from the manganese nodules by leaching them with aqueous base solutions such as aqueous solutions of sodium hydroxide or sodium carbonate.

These solutions should preferably have a pH of at least 10. The leaching with the aqueous base solutions can be carried out at ambient temperatures or at the boiling point of the solution. The demetalation reaction is carried out substantially as fully described below in U.S. Patent 3,813,331.

The catalyst, after being employed in the demetalation reaction and having become catalytically deactivated, or spent, can be treated for the recovery of valuable metals. Thus, the catalyst, after becoming spent, may be treated to recover copper, nickel, molybdenum, or any two, or all three, of these metals. It may also be treated to recover therefrom any other component.

An advantage of the process resides in its economy with respect to hydrogen consumption. During the demetalation reaction, hydrogen is consumed and the consumption of the hydrogen adds to the cost of demetalation. Thus, reduction in the consumption of the hydrogen is economically desirable. Prior processes have often required consumption of hydrogen in amounts between about 450 and 1,000 cubic feet per barrel of hydrocarbon charge stock.

As compared to this, effective demetalation can be effected by the process in many instances with consumption of 50 to 300 cubic feet of hydrogen per barrel of hydrocarbon charge stock. It is believed that the reduced hydrogen consumption to a large extent is due to the sensitivity of the manganese nodules to the effects of sulfur. Manganese nodules, as well as other catalysts heretofore employed for the demetalation of hydrocarbon charge stocks, effect hydrogenation of molecules other than those containing metals.

Thus, the manganese nodules, as well as other demetalation catalysts, will effect hydrogenation of benzene rings, for example. This hydrogenation of molecules other than those containing metals therefore results in consumption of the hydrogen in addition to that related to demetalation and, from the standpoint of the desired demetalation, represents a waste of hydrogen. However, as contrasted with other demetalation catalysts, the manganese nodules, in the presence of sulfur, have essentially no activity for hydrogenating benzene and other aromatic molecules.

They will, however, hydrogenate olefins. Hydrocarbon charge stocks contain sulfur to a greater or lesser extent, and, regardless of whether the catalyst is subjected to a sulfiding pretreatment, the sulfur in the hydrocarbon charge stocks will effect a rapid sulfiding of the nodules. As a result, hydrogenation of the aromatic constituents of the charge stock is reduced with resulting reduction in the consumption of the hydrogen.

Whereas a rapid sulfiding of the nodules will occur from the sulfur in the hydrocarbon charge stocks, sulfiding pretreatment of the nodules, as previously described, is of value. It is believed that, under reducing conditions, a reduction of the metal oxides in the nodules can occur with consequent loss in surface area and diminished activity. The sulfides on the other hand are more stable to reduction. Thus, when the nodules are exposed to a reducing environment either before or during sulfiding as occurs when the sulfiding results from the sulfur in the charge stock, a prereduction or competitive reduction of the oxides can take place.

Demetalation of Hydrocarbon Charge Stocks

P.B. Weisz and A.J. Silvestri; U.S. Patent 3,813,331; May 28, 1974; assigned to Mobil Oil Corporation describe a process for the demetalation of hydrocarbon charge stocks containing metal impurities. The process comprises contacting the hydrocarbon charge stock with hydrogen and with a catalyst comprising salt water manganese nodules. These nodules have been previously washed with water having a temperature of at least 125°F and a total salts content of not more than 1,000 parts per million for a time sufficient to increase the accessible surface area of the nodules.

Various hydrocarbon charge stocks such as crude petroleum oils, topped crudes, heavy vacuum gas oils, shale oils, oils from tar sands, and other heavy hydrocarbon fractions such as residual fractions and distillates contain varying amounts of nonmetallic and metallic impurities. The nonmetallic impurities include nitrogen, sulfur, and oxygen and these exist in the form of various compounds and are often in relatively large quantities.

The most common metallic impurities include iron, nickel, and vanadium. However, other metallic impurities including copper, zinc, and sodium are often found in various hydrocarbon charge stocks and in widely varying amounts. The metallic impurities may occur in several different forms as metal oxides or sulfides which are easily removed by single processing techniques such as by filtration or by water-washing. However, the metal contaminants also occur in the form of relatively thermally stable organo-metallic complexes such as metal porphyrins and derivatives thereof along with complexes which are not completely identifiable and which are not so readily removed.

The presence of the metallic impurities in the hydrocarbon charge stocks is a source of difficulty in the processing of the charge stocks. The processing of the charge stock, whether the process is desulfurizing, cracking, reforming, isomerizing, or otherwise, is usually carried out in the presence of a catalyst and the metallic impurities tend to foul and inactivate the catalyst to an extent that may not be reversible.

Fouling and inactivation of the catalyst are particularly undesirable where the catalyst is relatively expensive, as, for example, where the active component of the catalyst is platinum. Regardless of the cost of the catalyst, fouling and inactivation add to the cost of the processing of the charge stock and therefore are desirably minimized.

Demetalation of the hydrocarbon charge stock can be effected by thermal processing of the charge stock. However, thermal processing results in conversion of an appreciable portion of the charge stock to coke and the portion of the charge stock converted to coke represents a loss of charge stock that desirably should be converted to a more economically valuable product or products. Moreover, by thermal processing, the metallic impurities tend to deposit in the coke with the result that the coke is less economically desirable than it would be in the absence of the metals.

Demetalation can also be effected by catalytic hydroprocessing of the charge stock. However, catalytic hydroprocessing results in the catalyst becoming fouled and inactivated by deposition of the metals on the catalyst. There is no convenient way of regenerating the catalyst and it ultimately must be discarded. Since these catalysts are relatively expensive, catalytic hydroprocessing to demetallize hydrocarbon charge stocks has suffered from adverse economics.

It has been found that an economical and effective demetalation of a hydrocarbon charge stock is obtained using manganese nodules. Manganese nodules are readily available in large quantities and are relatively inexpensive. Further, material derived from the nodules is capable of effectively removing the metallic impurities from a hydrocarbon charge stock.

Thus, whereas the material obtained from the manganese nodules becomes fouled and inactivated by the demetallizing process, the material is obtainable at such low cost that the fouled and inactivated material can be discarded without significant effect on the economics of the demetallizing process.

This process is predicated upon the discovery that the catalytic activity of salt water manganese nodules for demetalation, and desulfurization, can be improved by washing the nodules with water. The washing is carried out with water having a temperature of at least 125°F. Further, the water has a total salts content of not more than 1,000 parts per million.

This washing effects an increase in the accessible surface area of the nodules. By accessible surface area, is meant the surface area of the nodules which the hydrocarbon charge stock is able to contact. It also effects reduction in the chloride content of the nodules. It is believed that the improvement in the catalytic activity of the nodules is due to the increase in the accessible surface area. As stated, washing is carried out with water having a temperature of at least 125°F.

With this temperature, improvement in the catalytic activity of the salt water manganese nodule is obtained. Higher temperatures, however, are preferred. For example, the temperature of the water employed for washing is preferably about the boiling temperature, namely about 212°F. Still higher temperatures may be employed where the washing is carried out under higher conditions of pressure.

The water employed for washing has a total salts content of less than 1,000 parts per million. Satisfactory improvement in the catalytic activity of the nodules is obtained with water having this total salts content. For example, the water may have a total salts content of not more than 500 parts per million. Preferably, fresh water is employed, i.e., potable water. Distilled water may also be employed.

By washing is meant contacting the salt water manganese nodules with the water. With contacting, a diffusion of the water into the nodules occurs. Further, a diffusion of a solution of salts out of the nodules occurs. Thus, washing can be effected by immersing the manganese nodules in the washwater. Immersion may be effected one or more times employing new washwater for each immersion.

Preferably, in washing, the manganese nodules and the washwater are moved relative to each other. Thus, where washing is effected by immersion, the washwater may be stirred or otherwise agitated with the manganese nodules being stationary or both the washwater and the manganese nodules may be moved relative to each other by stirring sufficiently vigorously to move the manganese nodules and washwater may also be tumbled to effect relative movement as by rocking or otherwise effecting oscillating movement of the container holding the manganese nodules and the washwater.

Washing may also be effected by passing the water over the manganese nodules as by passing the water through a bed of the nodules. Washing may also be effected by immersing the manganese nodules in the washwater and subjecting the washwater to boiling. A particularly effective way of washing the manganese nodules is by Soxhlet extraction.

The time of washing should be sufficient to effect a significant increase in the accessible surface area of the manganese nodules. By significant increase is meant an increase sufficient to effect a measurable improvement in the catalytic activity of the nodules. The time of washing to effect any desired degree of increase in the accessible surface area of the nodules will depend upon the temperature and the total salt content of the water. It will also depend upon the volume of the washwater relative to the volume of the manganese nodules and the degree of movement of the washwater relative to the nodules.

With higher temperatures and lower total salt content, shorter times may be employed. Similarly with greater volumes and greater degree of movement, shorter times may be employed. Generally, any time of washing of at least five minutes will increase the accessible surface area of the nodules sufficiently to effect a measurable improvement in the catalytic activity of the nodules.

Prior or subsequent to the washing, the manganese nodules can be crushed and sized to obtain a desired particle size depending upon the type demetalation operation employed, for example, a fixed bed operation, an ebullition operation, or otherwise. The demetalation reaction is carried out by contacting the hydrocarbon charge stock simultaneously with the catalyst and with hydrogen.

The temperatures at which the reaction is carried out can be from about 650°
to 850°F. At the higher temperatures, a greater degree of demetalation occurs.
However, the temperature employed should not be so high as to effect an undesirable degree of alteration of the charge stock. Preferably, the temperatures
employed are in the range of 750° to 850°F.

The pressures at which the reaction is carried out can be from about 100 to
about 3,000 pounds per square inch gauge. Preferably, the pressures employed
are in the range of 500 to 2,000 pounds per square inch gauge. Where the reaction is carried out by passing the hydrocarbon charge stock through a bed of
the catalyst, the liquid hourly space velocity (LHSV) of the charge stock can
preferably be from 0.5 to 2 volumes of charge stock per volume of catalyst per
hour.

Hydrogen circulation is at rates of 5,000 to 10,000 standard cubic feet of hydrogen per barrel of hydrocarbon charge stock. The hydrocarbon charge stock
along with the hydrogen may be passed upwardly through a fixed bed of the
catalyst in an upflow reactor or may be passed downwardly through a fixed bed
of the catalyst in a downflow trickle-bed reactor. The reaction may also be carried out by passing the charge stock and the hydrogen through an ebullient bed
of the catalyst. The reaction may also be carried out by contacting the charge
stock, the hydrogen, and the catalyst in a batch reactor.

The process may be employed for the demetalation of any hydrocarbon charge
stock containing organo-metallic compounds. Ordinarily, these will be hydrocarbon charge stocks containing sufficient metal to cause difficulty in the processing, or other subsequent use, of the charge stocks.

Other subsequent use of the charge stocks can include burning of the charge
stock as fuel wherein the metals cause corrosion problems. These charge stocks
include whole crude petroleum oils, topped crude oils, residual oils, distillate
fractions, heavy vacuum gas oils, shale oils, oils from tar sands, and other heavy
hydrocarbon oils.

The process can be carried out in conjunction with subsequent steps of processing
of the hydrocarbon charge stock. For example, the hydrocarbon charge stock
can be subsequently processed for removal of sulfur and/or nitrogen. Further,
for example, the hydrocarbon charge stock can be subsequently processed by
catalytic cracking.

Hydrocarbon Conversions

J.N. Miale; U.S. Patent 3,825,486; July 23, 1974; assigned to Mobil Oil Corp.
describes the conversion of hydrocarbons employing a catalyst comprising a base-
exchanged, calcined manganese silicate-containing mineral in the form of a manganese nodule. The mineral has hydrogen ions bonded thereto in an amount of
at least 0.01 gram per 100 grams of the mineral.

Manganese nodules provide an exceptionally good source of manganese silicate.
These manganese nodules can be treated to replace a certain portion of the alkali
or alkaline earth metal cations with hydrogen ions. The manganese nodules
utilized as a catalyst contain not only manganese, but several other important

metals including iron, cobalt, nickel and copper. The manganese nodules are found on the floor of oceans and are particularly abundant in the Pacific Ocean. The nodules are characterized by a large surface area, i.e., in excess of 200 m^2 per gram. The manganese nodules have a wide variety of shapes but most often they look like potatoes. Their color varies from earthy black to brown depending upon their relative manganese and iron content.

The nodules are porous and light, having an average specific gravity of about 2.4. Generally they range from 1 to 9 inches in diameter but extend up to considerably larger sizes approximating 4 feet in length and 3 feet in diameter and weighing as much as 1,700 pounds.

The nodules utilized as catalysts may be of any suitable particle size useful in the particular conversion operation. Thus, the particle size of the manganese nodules may vary from fairly large pieces down to and including powdered material useful in a fluidized catalytic operation. Manganese nodules, it should be understood, can occur in several different salt forms. They can occur as silicates and they can occur as manganese dioxide.

The manganese nodules contemplated for treatment by this process are manganese silicate nodules. These can be recovered from naturally occurring deposits, especially from deposits on the floors of the Atlantic Ocean and Pacific Ocean. The various manganese silicate nodules assay over a considerably wide range and include various metals including nickel, copper, cobalt, zinc and molybdenum. Some of the minerals in the nodules have no names, because the exact crystalline structure has never been encountered before.

Generally speaking, manganese nodules in the form of a manganese silicate are characterized by having the broad assay range as follows: 11.4 to 90% MnO_2, 2.8 to 42.3% SiO_2, 0.8 to 44.1% Fe_2O_3, 3.7 to 29.7% water and 0.3 to 12.8% alumina. It will be seen that the manganese dioxide content of the manganese silicate nodule can vary over a considerably wide range. This is due to the fact that these nodules are not synthesized but are recovered and mined generally from ocean floors.

Natural manganese nodules exhibit a catalytic action in the conversion of hydrocarbons. When, however, it is converted to the hydrogen form, in which at least a portion of its alkali metal and/or alkaline earth metal cations are replaced by hydrogen ions, its catalytic action is considerably increased.

To effect such conversion, the nodules are treated with a fluid medium containing hydrogen ion precursors to give a composite which, after calcination, contains hydrogen ions. Composite is the term applied to the mineral after a portion, at least, of its alkali and/or alkaline earth metal cations has been replaced by hydrogen ion precursors.

The step involved is base exchange, followed by calcination, and the hydrogen ions are bonded to the manganese nodules, thereby forming the composite. The latter is strongly acidic as a result. To illustrate the treatment, the nodules can be arranged in the form of a fixed bed, and the fluid medium in the form of an aqueous solution is passed slowly through the bed at room temperature and atmospheric pressure for a time sufficient to substantially replace the alkali metal cations of the original nodules.

The aqueous solution is characterized by having a pH above that at which the nodules decompose, preferably above 4.5. When the treatment is finished, the resulting composite is washed with distilled water until the effluent washwater has a pH between 5 and 8.

The fluid medium is preferably aqueous. Polar solvents are useful and may be aqueous or nonaqueous. They include organic solvents which permit ionization of hydrogen-containing substances added thereto, and include cyclic and acyclic ethers such as dioxane, tetrahydrofuran, diethyl ether, diisopropyl ether, etc.

A preferred base-exchange procedure comprises treating the nodules with an aqueous solution of a compound which supplies hydrogen ion precursor, such as ammonium ion, washing the material as described, drying it at 100° to 300°F, and then heating it to a temperature below its decomposition temperature to convert the substituent ammonium ions to hydrogen ions. The concentration of ammonium compound in the base-exchange solution is usually up to 5% by weight.

Base exchange may be carried out at ambient temperatures and below to temperatures just below that at which the nodules decompose. Pressures may vary from subatmospheric to superatmospheric, and the duration of treatment is that sufficient to permit substantial replacement of alkali metal and/or alkaline earth metal cations.

At the conclusion of the step, the material is dried, as by heating to 100° to 300°F for a period of up to 10 to 20 hours. The dried material is then calcined in air at 1000°F for up to 20 or more hours. Calcining converts the ammonium or substituted ammonium ion to hydrogen ion. The resulting composite may have bonded thereto at least 0.01 gram, preferably 0.01 to 0.5 gram of hydrogen ions per 100 grams of composite.

If desired, the nodules, either before, during, or after base exchange, may be mixed in any desired way with a matrix, generally comprising an inorganic oxide of porous character, which can serve as a binder and, if suitably chosen, may serve as an auxiliary catalyst.

Appropriate matrixes include silica-alumina gel, silica gel, alumina gel, as well as gels of alumina-boria, silica-zirconia, silica-magnesia, and the like. Natural clays are useful, such as kaolin, attapulgite, kaolinite, bentonite, montmorillonite, etc., and if desired, the clays may be calcined or chemically treated as with an acid or an alkali.

OFFSHORE MINING

UNITED KINGDOM MARINE MINING INDUSTRY

The United Kingdom marine sand and gravel mining industry is the largest and most advanced offshore mining operation of its type in the world, supplying an increasingly large portion of the concrete aggregate for the construction industry of the United Kingdom (U.K.) and bordering nations on the European Continent.

Some 32 different companies are operating more than 75 sand and gravel sea dredgers in U.K. waters, with a total capital investment for dredgers alone approaching £40,000,000, or $100,000,000. With the exception of a few barge-mounted grabs, these vessels are principally suction hopper dredgers, mostly with trailing pipes and many with self-discharging systems. Most dredging operations are conducted on a 24-hour cycle, taking advantage of the high tides for approaching shallow-water discharge points.

Annual production of sea-won aggregate in 1970 was approximately 14 million tons, or about 13% of the total U.K. production, and is steadily increasing. Reliable estimates indicate that the annual offshore production in U.K. waters will approach 20 million tons in 1971, with about 12 to 13 million tons landed in the U.K. and the balance exported to the Continent. Value of the unprocessed aggregate ranges from about 12 shillings ($1.44) per ton in the U.K., to about 22 to 26 shillings ($2.64 to $3.12) per ton at more distant points on the Continent.

Approximately 80 different dredge sites fall within six principal offshore dredging areas bordering England and much of Scotland. They supply aggregate to as many as 80 and perhaps more than 100 different discharge points on the U.K. coast, plus sizable quantities exported to bordering coastal nations of Western Europe. Shoreside treatment plants, primarily for washing and sizing the aggregate, are operated at most discharge points. Shipboard processing facilities, although few, are highly advanced.

The U.K. sand and gravel industry, as a whole, is the largest supplier of raw materials to that country's construction industry, most of its output being used in the form of fine and coarse aggregate for concrete. During the past 50 years it has developed from a relatively small industry with an output of 2,500,000 tons in 1919 into a major industry with production now amounting to some 110,000,000 tons per year. Of this amount approximately 14,000,000 tons per year, approximately 13% of the total U.K. sand and gravel production, comes from offshore production areas.

The growth of the U.K. sand and gravel industry has consistently exceeded the national population growth rate, in terms of per capita consumption. 1971 output, from both land and sea, corresponds to a per capita demand of over 2 tons per year. Projections made by the Sand and Gravel Association of Great Britain (SAGA) indicate a U.K. sand and gravel demand of 200 million tons by 1975 and close to 300 million tons shortly thereafter.

Based on these estimates and projected population figures, per capita consumption of sand and gravel in the U.K. is expected to increase sharply to 4 tons by 1975 and more than 5 tons soon thereafter. For comparison, U.S. per capita consumption for sand and gravel was 1.82 tons in 1940, 4.75 tons in 1966, and now exceeds 5 tons. Long-range forecasts indicate that demand for sand and gravel in the U.K. will increase to approximately 2 billion tons by the year 2000.

The almost insatiable call for sand and gravel for concrete over the past 50 years has led to heavy demands on land-based resources near large development areas, with serious shortages developing in areas such as London and the southeast part of England.

The problem is most acute where nearby available reserves are rapidly becoming depleted, and where land-use conflicts and amenity costs impose an additional hardship on land-based operations. This, coupled with rapidly expanding urbanization, growth in population, and increased gravel demands for the construction industry is among the factors which influenced the recent expansion in the U.K. offshore sand and gravel mining industry.

In the last 5 years or so, existing dredging companies have increased their production capacities many-fold and a number of sand and gravel operators who formerly had only land-based resources have extended their operation to marine dredging.

Development of the United Kingdom offshore sand and gravel industry has largely followed the pattern set earlier by land-based operations. In the beginning, smaller companies worked richer deposits closer to the market, followed by larger companies or small company consortiums entering the field and working lower-grade deposits farther from the market.

This general trend most likely will continue with offshore dredgers extending their sphere of operations to greater distances offshore and to greater water depths. Deposits worked in deeper waters, however, may not be lower grade; in fact, indications at this time are that many of them may be even richer.

Financial incentives once offered to the United Kingdom offshore sand and gravel industry in the form of a 20% government grant for building dredgers is no longer considered necessary, and this program has been terminated.

Dredge Types and Operations

More than 75 vessels, operated by about 32 different companies, make up the fleet used for dredging sand and gravel from the waters that surround the United Kingdom. Most of the ships have been specifically built for production of aggregate, but a few converted vessels exist.

The total capital investment of the U.K. dredging fleet is approaching $100 million, growing larger each year. A considerable investment also is required for shore-based support facilities including wharves, stockpiling and processing facilities, treatment plants, and vehicles.

Cargo capacities of sea dredgers vary from about 500 to near 10,000 tons. Most dredgers, primarily those built since 1968, can hold between 1,000 and 4,000 tons. The length generally ranges from about 150 to 250 feet, though the newest dredgers are mostly in the 250 to 350-foot class. Some of the largest seagoing dredgers operating off the U.K. coast are of Dutch design and are being operated by Dutch firms. In 1968 the average gross tonnage of sand and gravel dredgers operating in the U.K. was 800 tons. Then, during the 1969 to 1970 growth period, the average cargo capacity of newly built dredgers rose sharply to about 3,500 tons, and those planned or being built during early 1971 have capacities between 2,500 and 10,000 tons.

It was announced in early 1971 that a Dutch sea dredging company having a U.K. subsidiary was building a 10,000-ton capacity sand and gravel dredger. Other newly constructed dredgers include gross tonnages of 4,980, 3,000, 2,880, and 2,600. Also, in early 1971, Westminister Gravels, Ltd., placed an order for a trailing-suction gravel hopper dredger having a capacity of 8,000 tons and capable of self-discharging at the rate of 2,000 tons per hour. This vessel, having an OAL of 350 feet and the capability to dredge in 118 feet of water, is designed to screen off gravel and to discharge sand and coarse stone over the side.

Except for a few old barge-counted grabs, the U.K. sea dredging fleet is made up of suction hopper dredgers. Some operate with a forward-leading pipe, but most of the newer vessels trail the pipe. The trend in new designs is exclusively for trailing-type hopper dredgers.

With the heavy demand for gravel aggregate rather than sand, particularly in the Greater London area and on the Continent, dredge designs during the past few years have been developed for selective recovery of gravel and discarding of sand at sea. One major company has two such gravel dredgers, one operating out of London and the other out of Liverpool, and has plans for adding others to its fleet in the near future.

Gravel dredgers typically recover a cargo composed of more than 90% gravel with less than 10% sand from deposits containing about 30% gravel and 70% sand. Obviously in such operations, 2 to 3 tons of sand are discharged over the side for every ton of gravel recovered.

In fulfilling the present demand for gravel, dredgers of this type provide for improved operational efficiencies and obvious economic advantages over conventional sand and gravel dredgers in that they can operate on near-shore, lower-grade deposits located closer to the point of discharge.

Since U.K. gravel dredgers incorporate some of the latest and most advanced innovations in sea dredging for aggregate, they are thought to be worthy of special mention. Although they are equipped with trailing pipes, dredging is most commonly accomplished while at anchor.

Based on the heavy demand for gravel aggregate as opposed to sand, coupled with an increasing demand for immediate delivery of specific-sized gravel products or special aggregate mixes to distant ports, particularly on the European mainland, specialized large-capacity dredgers of advanced design are now being developed in increasing numbers.

One of the largest and most advanced suction dredgers presently in operation is a converted 8,000 DWT tanker El Flamingo. This dredger, having an overall length of 423 feet and a cargo capacity in excess of 7,000 tons, is equipped with a highly automated shipboard treatment plant capable of producing a wide range of washed and sized aggregate products at sea, then transporting the cargo to distant ports and unloading a desired sized product or special mix with an automated self-discharging system.

Representatives of U.K. industry look to the El Flamingo as a prime example of highly advanced dredge designs expected to characterize the U.K. sand and gravel sea-dredging fleet of the future.

Present sand and gravel sea-dredging standards demand modern, high-volume operations extending further and further seaward in dredgers of increasing capacities to mine larger reserves at greater depths.

It is no longer uncommon for a dredger to go to sea and recover 2,500 tons of sand and gravel in water 90 to 100 feet deep through a 24 to 36 inch diameter pipe, steam to port and transfer her cargo to shore with self-discharging equipment, then repeat the same operation, all in 24 hours.

Most dredging operations are governed by the tides, operating on a 24-hour cycle and taking advantage of the high tides for leaving and returning to normally shallow-water cargo discharge points in estuaries or far up rivers. Following cargo discharge, which normally takes about 1 to 2 hours, the dredger immediately heads back out to sea during the high tide cycle to take on another load. Most dredgers recover mixed sand and gravel, as dredged, but some keep only gravel, rejecting the sand at sea.

The normal ratio of sand to gravel as mined is approximately 70/30, while ideal dredge material is considered to be 40% sand and 60% gravel. In areas where such a ratio cannot be obtained, ship-board screening is often employed.

The majority of deposits being worked are relatively near shore (1 to 20 miles), comparatively close to market (in the U.K., a few to about 100 miles), and generally are in 60 to 100 feet of water. Deposit thickness normally ranges from 3 to 30 feet.

Present dredging techniques used for winning sand and gravel off the coast of the United Kingdom generally permit recovery to about 100-foot water depths. Operators naturally prefer shallow waters of about 30 to 40 feet, but are extending present operations into deeper and deeper waters.

In almost all instances, pumping is done by a contrifugal pump mounted in the ship below the waterline. Provided that there is adequate power and the pump is large enough, suction dredging appears to work staisfactorily by this method in water as deep as about 120 feet or more, but most present dredge ships can only reach to about 90 feet.

Water depth is important, since once it exceeds about 100 feet, the conventional suction pump is not considered efficient, and either jet-assisted suction pumps or pure jet pumps must be used.

In 1970 at least one large company is trying an injection pump, which reportedly can lift material from water 150 feet deep. One company reportedly is trying jet-assist suction pumps on one of its six dredgers, but some problems apparently have been encountered, principally clogging of the 10-inch suction pipes.

With good quality gravels being found at greater water depths, ranging up to 200 or 300 feet, the trend is toward dredging in deeper waters. Dredgers in the North Sea are now extending operations to deposits lying under 100 to 150 feet of water.

Also in the North Sea, where deposits are generally shallow and resting on a clay bedrock, trailer dredgers are normally required. However, a number of dredgers operating in this area are equipped with forward-facing suction pipes, presenting the danger that clay will be pumped and the cargo degraded.

On the newer dredgers, all phases of the dredging operation are normally remotely controlled from the wheelhouse or bridge, where an elaborate system of electronic or hydraulic control panels is centrally located.

From this vantage point, the forward and aft pipe gantries are extended, and the suction pipe lowered over the side in preparation for dredging. Then the attitude and location of the pipe during dredging is very rigidly controlled from this point, with a view toward optimum recovery of good quality material.

Dredging technology has generally kept pace with expanding demand, but unloading techniques have in some cases lagged behind in efficiency and cost. A number of different types of unloading methods have been tried, but only a few have met with real success in terms of efficiency and economy.

Since a considerable sum of money is invested in the dredge ship, it is important that the dredger not spend half her working life unloading alongside a dock. With larger dredge ships, the trend is away from conventional grab off-loading, which is reliable but very costly, for it may take 3 to 4 hours to empty a 2,000 ton ship that lifted its cargo in about 2 hours. Dock-mounted grabs, or clamshells, are still, however, a common means of unloading.

Market Trends

With increased population growth and increasingly large demands for new construction in the United Kingdom and on the Continent, demands for sea-dredged
aggregate are expected to increase substantially over the next few years, with
the outlook of total U.K. sand and gravel production coming from the sea in
about 10 to 15 years.

Contributing substantially to this trend is that land-based reserves are becoming
increasingly scarce and most costly to exploit as a result of raw material depletion and problems associated with urbanization, namely increased land costs,
land-use conflicts, and a variety of associated political, social, and environmental
amenity restraints.

In contrast, offshore reserves, which occur abundantly off the U.K. coast, are
relatively unaffected by such land-related problems, at least the ones that re-
flect increased product costs.

Projections made by the Sand and Gravel Association of Great Britain indicate
that future U.K. demands for aggregate will be at least 200,000,000 tons by
1975 and possibly 300,000,000 tons within a few years thereafter. Based on
comparable projected population figures, the projections would be equivalent to
4 or 5 tons per person, which would appear to be a reasonable figure.

As onshore sand and gravel supplies further diminish and access to minable reserves becomes more difficult, the unit price of the material is expected to rise
accordingly and the demand for sea-dredged material to become increasingly more
intense.

The increased price and favorable marketing climate for sea-dredged materials
thus should enable operators to search further afield in deeper waters. As steam-
ing distances become longer and reserves in deeper waters become more in demand, increasingly efficient dredging equipment and practice will be employed,
reflecting higher profit margins.

The outlook is for larger dredgers in excess of 10,000-ton capacity with treatment plants on board, and perhaps even larger ships having the capability to
carry the bulk material to a central dispersal point from which it can be distributed to the market by smaller ships after cleaning and sizing.

Unless there is a complete reversal of political, social, and economic factors
governing exploitation of land reserves, all indications are that the U.K. marine
aggregate industry and the market it supplies will continue the rapid growth
they have experienced in recent years.

With an inevitable hardening of prices for land-won aggregates, the profit margin on sea-dredged material should increase provided that technologic advances
keep pace.

Environmental Considerations

Environmental problems and restraints associated with sea dredging for sand
and gravel off the coast of the United Kingdom primarily relate to fisheries

ecology, coastal erosion, and navigation. The major objections to sea dredging appear to be based largely on fears created by a lack of knowledge rather than on scientific fact generated through research.

Certain fisheries, coastal erosion, and navigational problems are known to exist, but British authorities feel that at least some may be due to natural causes, or conditions other than sea dredging. In some instances, evidences strongly indicate that sea dredging may actually enhance the environment.

Until such time that comprehensive environment monitoring/prediction programs can be initiated, the true nature and extent of environmental disturbances resulting from sea dredging will remain uncertain and subject only to speculation.

UNDERWATER EXPLOITATION OF MINERAL RESOURCES—U.S.S.R.

The following information is taken from JPRS: 47,700. Many years of experience with the exploitation of deposits of rutile and zircon near the seashore of Australia have shown that the cost price of concentrates obtained by hydraulic underwater exploitation is several times lower than the cost price of concentrates obtained by exploitation of the usual deposits on land.

During a number of years on the shores of the Republic of South Africa at the mouth of the Orange River, diamonds were excavated with an air lift dredger from the floor of the sea at a depth of up to 100 meters. According to the data of American companies the cost price of tin obtained (in Malaya and Alaska) from sea deposits is 45% lower compared to the cost price of the same minerals extracted on land.

On the coast of Japan magnetite sands are excavated with a grab from a depth of 27.5 meters. This costs only half as much as the extraction of iron ore on land. The annual output of the enterprise is seven million tons of sands.

According to data of foreign experiences, a generalization of the work in underwater exploitation of mineral resources shows its economic effectiveness.

	Conventional Exploitation	Underwater Exploitation
Cost price of the production of one carat of diamonds, in shillings	120	30.6
Specific capital investments, in percent	100	15–20
Cost price of the production of cassiterite, in cents	82	25–37
Extraction of Magnetite-Vanadium Sands		
Cost price of the production of one ton of mineral, in dollars per ton	8–10	5
Capital investments for one ton of mineral, in dollars	70	40

(continued)

	Conventional Exploitation	Underwater Exploitation
Cost price of the production of one ton of nickel-manganic concretions, in dollars per ton	14	2.29–3.5
Specific capital investiments for the production of one ton, in dollars	12	3

The high technological-economic indices of underwater exploitation of minerals are achieved because in many cases there is no need for uncovering, the volumes of the capital trenches are much smaller or there is no need for them at all.

Furthermore, it is not necessary to build dumps and places for residue, the preparatory works within the outlined deposits are minimal, and there is no need to build spur-tracks. The excavation of deposits with sea suction dredgers or drags permits exploitation of mineral resources without special expensive work.

The power-equipped dredgers permit excavation independent of electrical power lines and power sources. Therefore the exploitation of underwater deposits of minerals can be done in a very short time and with considerably lower specific capital investments than are necessary for construction of mining enterprises under usual conditions.

The systematization of underwater sites of mineral deposits showed that there are a number of regions in the seas of the Soviet Union which might be ore-bearing.

	Possible Mineral Resources
Baltic Sea	Ilmenite, rutile, zirconium
Black Sea	Ilmenite, rutile, zircon
Laptev Sea, East Siberian Sea	Cassiterite, gold
The Sea of Japan	Cassiterite, gold, zircon, vanadium, rutile, ilmenite
Sakhalin Island	Cassiterite, ilmenite, rutile, zircon
The Kuril Islands	Titanomagnetite, vanadium
The Kamchatka Peninsula	Gold

In 1966 the U.S.S.R. Ministry of Nonferrous Metallurgy conducted experimental underwater exploitation of ilmenite-rutile-zirconium-containing sands in the Baltic Sea.

Experimental-industrial exploitation was conducted with a sea dredger with automatic removal and dragging suckers. It has a productivity of 11,000 cubic meters of pulp per hour, a hold capacity of 1,180 cubic meters, a draught of 4.2 meters, and a gear speed of 11.3 knots.

The field of the experiment was 300 x 100 meters. Outlined by buoys, it was 600 meters from the shore. The ocean floor had a sandy slope, gently dropping at a depth of five to eight meters. Before the dredger began to work, divers investigated the floor and measured the primary depths with a sonic depth finder.

On the outlined field the mining substance, forming a layer of up to 30 cm, was excavated by moving the dredger parallel to the shore in separate lanes 16 meters wide and one kilometer long. The hold was filled with mining substance in an average of two hours. After that the dredger took the mining substance to the harbor on a place ashore.

The place for receiving the mining substance was equipped with a mooring specially constructed out of mining bollards, and a shore pulp line with openings for emission. The ocean part of the field was protected by a groove. Mining tiles were used to extinguish the pulp energy. For transporting the pulp to the shore a transition mechanism with a diameter of 700 mm was installed on the dredger and the scow.

After fastening to the mooring line the pulp moved through the transition mechanism into the pulp line (diameter 500 mm) and was placed through special outlets 150 mm in diameter on the receiving field where a concentrating unit was installed. During the experimental and prospecting work approximately 200,000 cubic meters of sands were processed by the dredger, and 47,000 tons of sands were washed ashore.

The processing of ilmenite-rutile-zirconium sands showed that by using powerful hydraulic dredgers a complete excavation can be done without causing impoverishment of the mineral.

During the exploitation for experimental and prospecting purposes, the achievement of a higher efficiency of the excavation of the sands by application of coagulators was studied.

It was established that by using polyacrylamide the productivity of the dredger increases by 25 to 30%, and labor productivity increases by 25%. The expenses for the coagulators are less than 0.001 ruble per one cubic meter of sand. At the same time the cost of extracting one cubic meter of sand is lower. Below are the basic results of the research.

	Recommended Value
Nonimpairing speed of suction, m/sec	4.5–5.2
Optimal speed of motion on the working field, km/hr	3.4
Thickness of the excavated layer, meter	Up to 0.3
Sizes of lanes, meters	16.0
Length of the work front, meters	800–1,000
Maximal wind force	5.0
Optimal capacity of the hold, m^3	450
Coefficient of utilization of working time	0.68

(continued)

	Recommended Value
Method of concentrating the mineral substance in the process of silting	By application of polyacrylamide
Consumption of polyacrylamide per 1 m³ of mineral substance, grams	2.5

The research showed that it is expedient to deposit the concentration residue without special installations in the open sea at a minimal distance from the work field. Considering that the current drifting is from south to north, the residue has to be deposited north of the field of processing.

Studies of the extraction of rough concentrate on concentration tables, cone-shaped separators, and by flotation showed that it is not very effective to concentrate sea sands on these apparatuses. The use of electrical separation during the input of the collective concentrate resulted in exact separation of the minerals.

90 to 95% of the ilmenite was extracted into the conductors, zircon went almost completely over into the nonconductors where rutile and leucoxene could also be found. By using magnetic separation, ilmenite concentrate, corresponding to technological conditions, was obtained from the conductors. Because of the insufficient weight of the sample no zircon or rutile concentrates were obtained.

Experimental concentration equipment was installed in the port. It consisted of three flowing concentrates, one hydraulic cyclone separator, two sand and two water pumps. The excavator with the help of the PTS-2 belt feeder guaranteed an even feeding of the raw material into the sump of the sand pump. By concentrating the sands, 60 tons of rough concentrate were obtained. The monoconcentrates gained under laboratory conditions meet the required standards.

Calculations have shown that the cost of concentrates gained in underwater exploitation will be two to two and a half times lower than the cost of concentrates gained from deposits on land. Capital investments will also be considerably lower.

PROPRIETARY DREDGING PROCESSES

Mined Material Conveyed to Vertical Floating Vehicle

M.G. Krutein; U.S. Patent 3,438,142; April 15, 1969 describes a sea mining method and apparatus wherein desired minerals lying at the sea bottom are first conveyed from the sea bottom to the submersed portion of an elongate vessel floating substantially vertically in the sea and are then conveyed vertically from the submersed portion of the vessel to the surface of the sea.

The apparatus includes an elongate vessel having an extended tank portion sealed for submersion over a major portion of its length, ballast means for shifting the center of gravity of the vessel between a transport position with the sealed tank lying substantially horizontally in the sea and a mining position with the vessel and tank floating in a vertical position. In mining position the major portion of the sealed tank is submersed in the sea, and minerals are brought to the vessel by dredging or conveying equipment mounted on the furthest submersed portion of the vessel.

The dredged matter is brought into the submersed portion of the sealed tank. There the matter is dewatered and the desired minerals conveyed upwardly through the tank to the surface of the sea. The minerals are then preferably transported to an underwater stockpile where they can be reclaimed as desired.

The dredging vessel is designed to float in a vertical position during dredging or mining operations so that with the major portion of the vessel submerged the vessel is affected to a minimum extent by wave action adjacent the surface of the sea. This mining vessel is highly stable in vertical position thereby providing great stability and thus maneuverability to the actual equipment that picks up and moves the mined minerals.

The stability of the mining vessel permits ocean mining at virtually any depth desired since mining equipment extending to any depth can be attached to this vertically floating vessel. In one example, the cross-sectional area of the mining or dredging vessel is reduced or suitably shaped over the region of heavy wave action to maintain the effects of wave motion on the dredging vessel at a minimum.

With the mining or dredging equipment which conveys the minerals from the sea bottom to the vessel located on the submerged end of the vessel, the required length of the dredging or conveying equipment is effectively reduced by the submerged length of the vessel. This construction provides better maneuverability for the conveying equipment due to the shorter length of mining equipment in mining operations where the distance between the lower submersed end of the vessel and the sea bottom is small.

The dredging vessel is powered to move through the sea at selectible low rates of speed whereby the dredging operation can take place during patterned slow traverses across a mineral deposit lying on the sea bottom for maximum recovery of the desired minerals.

Hydraulic conveying equipment is used to convey the minerals from the sea bottom into a portion of the dredge vessel which is located far beneath the surface of the sea but maintained at atmospheric pressure due to communication with atmosphere upwardly through the vessel. This construction permits the utilization of the difference in pressure at the bottom of the sea and at the location within the dredging vessel to convey the minerals into the vessel.

After the minerals have been passed into the vessel and separated from the conveying hydraulic fluid or water, the water can then be conveniently pumped out of the vessel. This arrangement eliminates the amount of work required to move the minerals from the sea bottom to the level of the lower submersed portion of the vessel.

The material recovered from the sea bottom and conveyed into the dredge or vessel is separated as to desired and undesired minerals. Then the desired minerals can be conveyed vertically within the vessel to the surface of the sea and the undesired minerals exhausted from the vessel back into the sea. This arrangement reduces the amount of work that would otherwise be required, since it is unnecessary to convey undesired minerals the submersed length of the vessel.

The minerals recovered from the sea or even minerals transported across the sea to a processing plant are stockpiled and blended underneath the surface of the sea.

A number of loads of minerals are transported over the surface of the sea and each dumped in shallow water in a predetermined position of a pattern on the sea surface with the various loads evenly distributed over the pattern and a plurality of loads at substantially every position so that an underwater stockpile with a pattern of adjacent and superimposed mineral load layers is produced on the sea bottom.

The minerals in a number of different loads are simultaneously reclaimed and blended as desired from the underwater stockpile by cutting across and through load layers with dredging equipment.

By stockpiling raw materials under water, docking time for the barge is completely avoided so maximum advantage can be taken of good weather conditions for effectuating the mining and transporting operations over the sea. Additionally, the likelihood of damage to the barge and berth as is often caused during docking operations is avoided.

Furthermore, the cost of the offshore area on which underwater stockpiles are established is far less than that for conventional stockpiles positioned on dry land, and the cost of unloading the mineral stacking equipment is avoided.

One method and apparatus for reclaiming minerals from underwater stockpiles is a dredging vessel provided with fore and aft anchor assemblies for locating anchors at two opposing sides of the underwater stockpile and port and starboard anchor assemblies for locating anchors at the two other sides of the underwater stockpile.

Means are provided for moving the dredge vessel between the fore and aft anchor assemblies and the port and starboard anchor assemblies to cause the dredging apparatus on the dredging vessel to sweep arcs across the underwater stockpile for picking up or reclaiming minerals in a plurality of loads thereby effectively to blend the material as it is reclaimed.

A dredging apparatus at one end of a pontoon arm and means for rotating the pontoon arm about the opposite end to direct the dredging apparatus in arcs across the underwater stockpile are provided for reclaiming minerals from the underwater stockpile. The end of the pontoon arm that rotates is held against a barge which can be located between a pair of dolphins in a string of dolphins along one side of the underwater stockpile.

The dredged material is conveyed from the dredging apparatus along the pontoon arm to the barge and then to an adjacent land mass for conveyance to the processing plant. This assembly requires only the location of a plurality of dolphins along the side of the stockpile and the mooring of the barge to only two of the dolphins at any one time.

With this apparatus the reclaimed minerals picked up by the dredging apparatus can be passed through a separation stage either at the dredge end of the pontoon arm or at any position along the length of the pontoon arm so that by the time the minerals reach land they are separated and dewatered for use as desired.

Mining Head with Flexible Suspension Arrangement

N.B. Plutchak; U.S. Patent 3,543,422; December 1, 1970; assigned to The Bendix Corporation describes a mining head having a flexible suspension arrangement for mining material from the bottom of the ocean or from any location where the material can be converted into a slurry.

The head consists of a circular plate having a smoothly contoured circular passageway through its center connected with a large diameter hose and a bell-shaped member attached in close proximity to the central passageway to direct the flow into the hose. A manifold carries water at high pressure to a number of nozzles located around the edge of the plate, and flow from these nozzles puts the solids into suspension.

The resulting slurry is drawn into the central passageway by means of a pump which carries the slurry to the desired location. The head is suspended from a crane carried on a ship by means of a cable having a flexible link to take up vertical movement due to wave action. Flotation buoys are fastened to the hose to keep the loaded hose essentially neutrally buoyant irrespective of its length or the weight of the slurry carried.

The apparatus includes a mining head of very efficient design and also connections to the head including a supporting structure and conduits for delivering water and recovering material, usually sand, in a slurry. The head includes a large disc or plate with a plurality of nozzles for directing water or other fluid at high pressure aimed downwardly at the periphery of the plate.

Flow from these nozzles breaks the sand loose from its packed condition and converts the water and sand into a slurry which is confined below the plate member into a relatively restricted volume. At the same time, suction is produced at an intake at the center of the plate which draws the slurry into the intake and through a large diameter pipe or flexible hose to its desired location.

Since the sand is brought into and maintained in a state of suspension by the water jets directed into the surface, it is necessary to collect and transport this sand, and this is accomplished by creating a suction by means of a centrifugal pump at the head intake. This force must create sufficient velocity to keep as much as possible of the sand from settling out before it can be drawn into the intake port.

To keep this velocity as high as possible over a significant area, a plate or disc is used which is kept essentially parallel to the bottom (where the bottom is essentially flat), this making the volume over which the suction force operates limited severely as to height, although of larger area. Thus a substantially two-dimensional velocity field is attained rather than a three-dimensional velocity field, and the velocity falls off more nearly as the square root of the radius from the intake rather than as the cube root.

As a result, the velocities maintained under the plate are sufficient, for the most part, to keep the sand in suspension even near the outside edge where velocities are lowest. This velocity must be redirected at the intake from horizontal to vertical movement, and this is accomplished with minimum loss by using a bell-shaped intake which minimizes turbulence in effecting the transition into the intake hose.

Because of the weight of the mining head and the large diameter hose carrying the slurry, flotation buoys are typically attached to the hose to keep the hose approximately neutrally buoyant, and the entire assembly to an effective negative buoyancy of only a few hundred pounds (typically 200 to 300 pounds).

Although the mining apparatus might be located on shore, the usual installation is on a ship or large boat where complications are introduced because of the vertical movement resulting from wave action. The ship will normally carry pumps for pumping seawater into the smaller diameter hose to the nozzles and for creating the suction to pull the slurry into the large diameter hose and to transport it.

These hoses are supported by means of a cable carried on a crane extending over the surface. To compensate for the vertical movement of the ship, this cable includes a section which is pulled slack and paralleled by means of an accumulator, such as a flexible bungee member. The tension required to stretch the accumulator is somewhat less than the negative buoyance of the mining head assembly so that the head will remain close to the bottom surface despite the vertical movement of the ship.

Underwater Suction Head and Suction Pump

B.A. Boyle; U.S. Patent 3,563,607; February 16, 1971 describes a subaqueous mining machine which has an underwater suction head with a suction pump mounted on it. The pump is driven by a high-pressure air hose and a semi-buoyant discharge hose is used to conduct the pumped slurry to a shore-based treatment means. The underwater head can be steerable by means of jet control pumps which are part of it.

The miner comprises a suction head connected via a supply hose to a compressor located upon a tender barge adapted to float in the vicinity of the miner, but spaced from it to an extent which permits the miner to range about upon the seabed.

Preferably, a control cable which may be located within or without or alongside the supply hose, also carries electrical conductors from the barge to suitable illuminating means such as one or more electric lamps, disposed upon the head within the fairing together with a surveillance television camera, the monitor receiver of which may be located upon the barge or at a treatment plant to which the product hose is extended.

The head is also provided with one or more jet control pumps, with respective discharge ports, adapted to render the head steerable so that a monitor may observe the receiver screen and direct the head to a region upon the seabed which is clear of obstructions such as rocks and/or ensure that the head scans the seabed efficiently for the purpose of extracting ore.

If desired, an additional television camera and one or more electric lamps may be located behind a window to permit observation of the inlet duct of the product pump. It will be appreciated that because the head is a suction device, the reaction force exerted thereon by its air supply tends to bring the proboscis of the head automatically against the working face of the solid particulate material to be extracted.

The approximate position of the head is initially determined by the condition of buoyancy tanks, which have discharge ports. Furthermore, in view of the fact that the product pump causes the water and particulate material to be drawn from the working face by suction, there is little turbulence generated in the seawater in the vicinity of the working face, and hence little turbidity which might otherwise spoil the view of either of the television cameras.

If desired, the head may be subjected to an automatic scanning motion after a monitor has noted that the area of the seabed is generally acceptable for mining. Thus, the jet control pump may be automatically operable so that the head may swing from side to side in a series of progressively displaced arcs to an extent determined by the length of the supply hose (and hence the control cable) between itself and the barge, and also as determined by the amount of slack present in the product hose.

The barge may, of course, be moved from one reference position to another depending upon the amount of the slack in the product hose, so that the miner may scan in a number of different locations to an extent permitted by its supply hose while the barge is anchored. Alternatively, the barge, instead of being anchored in such reference positions, may be put under way so as to compensate for any winds and/or currents which may move it away from a particular reference position.

If desired, the compressor aboard the tender barge may be adapted to supply high-pressure air to the product pump and jet control pumps in such a manner that it does not exhaust back to atmospheric pressure but exhausts instead to air receivers adapted to give buoyancy to the barge, thus ensuring that the exhaust pressure is that due to the static head of immersion of the head. It has been found in tests that this last-mentioned expedient enables the energy requirements of the compressor to be reduced by approximately 30%.

Excavation of Sea Floor as an Extended Trench

A.J. Nelson; U.S. Patent 3,638,338; February 1, 1972 describes a method and apparatus arranged to excavate sea floors either as an extended area or trench either for recovery of material or preparation of a site. Apparatus is included to cope with undulations of the sea floor and vagrancies of the sea surface.

A method of establishing selected portions of the area to be worked and the mode of operation is developed to expedite work and prolong life of equipment. The reliance upon buoyant chambers to selectively support apparatus immediate to it minimizes stresses in the members comprising the assembly. Provisions are included to facilitate modification to the assembly even under adverse conditions.

A pair of cutters and a dredge pump mounted to a lowermost pontoon is immersed to penetrate the floor of a body of water employing an indicator to limit the depth of cut for the pendulous traverse over a selected floor area. Anchors are selectively embedded into the floor remote to the dredged area to which wires are connected extending from powered winches mounted on the lowermost pontoon.

The extended anchored wires are intermediately supported by a buoyed pendant so as to elevate those wires off the floor to avoid prolonged dragging and contaminating effect to them. Consequential with the catenary curve resulting from that elevation, a tension is developed creating an artificial demand of a slacked wire paid off one of a pair of identical winches at a greater rate than the opposite wire hauled-in to effect the pendulous swing established by a wire fixed to an anchor axially central with but remote to the selected area.

Each pendulous swing is at a radius shortened by a winch oriented in mounting with the two identical winches so that all wires lead from the winch as tangent lines direct to the anchor. The anchors are relocated by towing on a surfaced conduit pendant extending to a lower buoy of adjustable support capacity secured by a cable pendant to the anchor whereby the buoy lift capacity is increased to free the embedded anchor off the floor.

An articulative conduit in fluid communication between the pump and a delivery terminal on a surfaced service station is supported as a suspended tensioned array stabilized by immersed pontoons of selected and automatically adjusted support capacity responsive to changes encountered. An arrangement of hoists, structural provisions and utilization of pontoons common with the array facilitate the transfer of objects to and from the service station and the array to modify the dredging apparatus.

Compressed Air Operated Apparatus

A. Trondle; U.S. Patent 3,673,716; July 4, 1972 describes a compressed air operated apparatus for raising underwater deposits. The material to be raised is conveyed via a dredge pipe from the deposit to a separator in which the material is separated from the pressure medium and possibly from entrained portions of a liquid medium disposed above the deposit. The pressure medium is introduced at the lower end of the dredge pipe in such a manner that a partial vacuum is produced under the effect of which the material to be raised enters the dredge pipe.

The method has the advantage that the material is removed not only from the base but also from the wall of the deposit, a cavity being formed in the wall by removal of material therefrom and material slipping down as a result. This method is particularly advantageous when raising gravel from water-filled gravel pits.

Depending on the type of deposit, it may be advantageous to arrange the axis of the lower dredge pipe end substantially perpendicular to the surface of the deposit by the tension means and then to hold the inlet opening of the dredge pipe at a predetermined substantially constant distance from the deposit, which is preferably substantially equal to 1 to 1½ times the diameter of the inlet opening of the dredge pipe.

The loosening and removal of the material may be substantially promoted by cutting tools which preferably rotate concentrically to the inlet opening of the dredge pipe.

A substantial improvement in the output and material removal results if the pressure medium is introduced into the dredge pipe with a rotational component. The rotation of the pressure medium effects a turbulence in the region in front of the inlet opening of the dredge pipe which whirls up the deposit in this region.

In this manner a segregation of the material raised is very effectively avoided and material in front and lateral of the inlet opening of the dredge pipe is also loosened. Conveniently, the pressure medium is made to rotate about the dredge pipe axis.

The pressure medium is supplied at the inlet end of the dredge pipe in the form of a flow which is of annular cross section, moves along the inner surface of the dredge pipe and is substantially uniformly distributed over the periphery.

It may be advantageous to subdivide the pressure medium flow in the peripheral direction to obtain a more uniform bubble formation. The pressure medium may be guided in the injection nozzle by guide fins extending in the direction of flow to ensure a supply uniformly distributed over the periphery.

A floating free of the inlet opening of the dredge pipe may be effected in that pressure medium is conducted intermittently against the conveying direction. For this purpose, the dredge pipe is connected in the upper region of the conduits via a connecting conduit which may be shut off by a valve to the pressure medium conduit.

Furthermore, a shut-off valve is provided in the dredge pipe above the opening of the connecting conduit and in the pressure medium conduit below the opening of the connecting conduit. Normally, the valve in the connecting conduit is closed whereas the valves in the dredge pipe and in the pressure medium conduit are normally open. If the pressure medium is to be diverted to the lower end of the dredge pipe, the valve in the connecting conduit is opened and the two other valves closed.

Mining of Sand and Gravel Deposits Containing Gold Particles

It is generally known that gold and associated heavy minerals are widely distributed in the sand and gravel sediments of submerged beaches and drowned river valleys of the continental shelf areas. Local concentrations of gold have been identified in surface sediments in the continental shelf off sourthern Oregon, and there is reason to believe that such concentrations exist off the shores of all five continents.

The process and apparatus described by *L.A. Lindelof; U.S. Patent 3,731,975; May 8, 1973; assigned to QVA Corporation* contemplates concentrating or beneficiating the gold-bearing sand and gravel deposits by using the action of gravity and the ambient subsurface ocean current to produce an initial separation of the gold particles from the sand and gravel deposits.

In this regard, the sand and gravel deposits containing the gold particles is continuously excavated by an undersea mobile excavating device and the excavated material is moved to a predetermined height by suitable conveyor means. The excavated sand and gravel deposit is continuously discharged at a predetermined height directly into the ambient ocean current.

A collection device, preferably a reciprocating screen type, is positioned below and downstream from the discharge point. The difference in elevation and the downstream spacing between the discharge point of the sand and gravel and the reciprocating screen collection device is so related to the magnitude of the ambient current as to cause the gold particles and the coarser sand and fine gravel under the influence of gravity and the ambient current to fall downstream upon the reciprocating screen receptacle.

Although the gold particles are of comparable weight to the coarser sand and fine gravel particles, the gold particles are smaller and the reciprocating screen will produce a further separation.

The fine sand particles will be moved downstream beyond the reciprocating screen receptacle under the influence of the ambient current and the action of gravity while the heavier larger particles will fall upstream upon the ocean floor with respect to the collection device. Thus the use of the ambient current as a means of initially separating gold particles from sand and gravel deposits makes possible the economic mining of these deposits.

The process and mobile apparatus for the undersea mining of sand and gravel deposits containing gold particles and associated heavy minerals such as platinum, magnetite, etc., include an excavating device which continuously excavates the sand and gravel material.

A conveyor receives the excavated sand and gravel and conveys the material to a discharge point which is located a predetermined distance above and a predetermined distance upstream from a screen-type collection device.

The predetermined difference in elevation and the downstream spacing between the collection device and the discharge point is so related to the ambient ocean current adjacent the ocean floor that separation of the sand and gravel discharged from the discharge point takes place as a result of the action of gravity and the subsurface ambient current acting on the material.

The gold and associated metal particles (platinum, magnetite, etc.) and sand and gravel particles which are slightly coarser than the gold particles will be moved downstream by the ambient ocean current as the material falls and will be received by the collection device, while the sand particles will be carried downstream beyond the collection device, and the more coarse materials will fall upstream with respect to the collection device.

The apparatus includes a self-propelled crawler-type bucket wheel excavator which includes a revolvable bucket wheel having a plurality of buckets or scoops which are circumferentially arranged so that as the bucket wheel is revolved, the sand and gravel deposits will be excavated and will be directed to an endless conveyor.

The revolvable bucket wheel is mounted at the outer end of an elongate boom which is connected at its inner end to the body of the excavator so that it may be shifted. The body is mounted for traversing movement along the sea floor by suitable surface engaging crawler assemblies.

The endless conveyor conveys the sand and gravel deposits excavated by the revolvable bucket wheel to a suitable guide-type hopper device positioned adjacent the discharge end of the endless conveyor. The sand and gravel material are then conveyed to an elevator-type conveyor to a predetermined height where the material is discharged.

Thus the elevator-type conveyor has one end portion thereof which receives the sand and gravel material from the endless conveyor and has a discharge end or discharge point from which the sand and gravel material is discharged. The elevator conveyor is vertically adjustable so that the vertical spacing between the point of discharge of the conveyor and a collector device may be variously adjusted.

It is also desirable to properly orient the sand and gravel material discharged from the conveyor so that this material will be directed in its downward descent from a discharge point which is disposed in a vertical plane arranged substantially normal to the direction of the subsurface ambient ocean current.

To this end, a substantially flat inclined deflector device is provided and is positioned slightly below and adjacent the discharge end of the conveyor to receive the sand and gravel material discharged therefrom. The deflector device is inclined downwardly in the direction of the subsurface ambient current and may be provided with an upstanding flange affixed to the peripheral edges thereof.

Suitable direction control vanes may be affixed to the upper surface of the deflector device and these vanes may be preferably arranged so that they diverge towards the discharge edge thereof. It will be noted that the discharge edge of the deflector device is substantially straight and disposed in a vertical plane arranged substantially normal to the subsurface ambient ocean current.

Since the discharge edge or point of the deflector device is disposed in a vertical plane arranged substantially normal to the ambient current, the downstream spacing between the discharge point and the collection device may be more accurately determined. It is pointed out that the deflector device is also vertically adjustable with the conveyor so that the vertical spacing between the discharge point and the collector device may be readily adjusted.

The collection device is provided with a sectional reciprocating screen at its upper surface which is operable to produce further separation of the gold particles from the sand and gravel particles. The sectional screen includes a plurality of sections arranged side by side in which the mesh size of screen sections decreases. Additional screen sections could also be provided. The sectional reciprocating screen reciprocates so that the coarser material which will not pass through the screen section will be progressively conveyed along and discharged from the screen section.

For example, the gold particles will pass through the openings in the first screen section while the gravel and coarser sand will be conveyed longitudinally along the screen section and will be discharged therefrom.

These screen sections are so arranged that the smaller screen size is located further downstream from the larger screen size. Thus the heavier and larger particles will fall upon the larger mesh screen section.

One of the natural characteristics of the ocean is the essential condition which permits economic operation of the process and apparatus. This characteristic of the ocean is ambient subsurface current and it is necessary to determine the direction of the current, as well as the magnitude of the current for optimum results.

Generally speaking, the sand and gravel deposits constituting the ocean floor may be mined in any direction but it is essential that the excavated gold bearing sand and gravel deposits be discharged so that the material is influenced by the ambient current as material falls by action of gravity towards the collection device.

It will be seen that the collection device is positioned at a substantially lower level than the discharge point of the conveyor and is also positioned substantially downstream from the discharge point. The vertical spacing and downstream spacing between the discharge point and the collection device will vary according to the magnitude of the ambient current.

It has been found that the subsurface ambient current adjacent the ocean floor varies within the range of ¼ knot to approximately 4 knots. Although the vertical spacing between the discharge point of the deflector and the collection device will vary over a relatively large range, experimental evidence indicates that a desirable workable range for this vertical spacing is approximately 10 to 40 feet. It is also pointed out that the discharge point would be spaced vertically above the ocean floor a distance of 25 to 100 feet.

It has been found that the gold particles which are disseminated throughout the sand and gravel deposits of the Pacific continental shelf area, especially off the North American coast vary in size from particles which are of a size that will pass through a 10 mesh screen, but not a 20 mesh screen (plus 20) and those that are of a size that will pass through a 100 mesh screen (minus 100).

It is thought that probably the greatest amount of recoverable gold and associated mineral particles occur within the range of sizes between minus 40 and plus 100 and smaller. The weight of these individual particles has been generally calculated to be approximately 1.0072 grams for the minus 10, plus 20 particles, 0.003 grams for the minus 20, plus 40 particles, and 0.0005 grams for the minus 40, plus 100 particles.

The average descent time of gold particles in water, as well as other particles, can also be determined and this information must be known in order to properly interrelate the discharge point of the conveyor with respect to the collection device. Gold particles of a minus 10, plus 20 size have an average descent time of approximately 3.52 seconds in a 35-inch vertical column of water.

Similarly, gold particles of a minus 20, plus 40 size have an average descent time of approximately 3.88 seconds in water in the same column of water while particles of a minus 40, plus 100 size have an average descent time of approximately 5.44 seconds in the same column of water. Gravel, on the other hand, of a plus 10 size has an average descent time of approximately 4.13 seconds while the average descent time for sand depends upon the size of the sand particles.

In this regard, sand of a minus 10, plus 20 size has an average descent time of approximately 6.20 seconds, minus 20, plus 40 size sand particles have an average descent time of approximately 11.20 seconds, and sand particles of a minus 40, plus 100 size have an average descent time of approximately 23.20 seconds. The sand and gravel descent times were also based on a vertical column of water approximately 35 inches in height.

It has also been determined that the particular configuration of the gold particles affects the descent characteristics of the particle in water. In this regard, the heavier, bulky particles without being influenced by any ambient current will fall directly down without any spinning motion, while the flatter shaped particles while descending straight down at a slightly lower rate tend to spin rapidly about a vertical axis.

In every instance the gold particles descend faster than sand or gravel particles of the same approximate size. In a given screen section, smaller bulky shaped (fast descent) particles will be collected along with the larger particles for which the screen is sized.

It has been determined that the coarser sand and fine gravel of the ocean floor deposits will have the same general rate of descent as the rate of descent of the gold particles found in these deposits.

Thus, when the discharge point of the elevator conveyor is arranged in its predetermined relation with respect to the collector device, the gold particles within the range of minus 10, plus 20 to minus 100 (the size of particles normally expected in these deposits) and the sand and gravel particles within the range of plus 10 to approximately plus 60 will have the same general rate of descent and will fall upon the collector device and will be subjected to the action of the reciprocating sectional screen thereon.

Therefore the first screen section may be a 10 mesh size and the second and third screen sections may be progressively smaller. With this arrangement, the heavier but smaller gold particles will fall through the openings in the first screen section while the plus 10 gravel particles will be conveyed as tailings from the section.

The progressively smaller screen sizes will permit the same separation with respect to those sand, gravel and gold particles carried further downstream by the ambient current and the action of gravity.

When the direction of the ambient ocean current as well as the magnitude thereof is determined, the vertical spacing between the discharge point for the excavated sand and gravel deposits and the collection device may be determined and the downstream spacing between the discharge point and the collection device may also be determined.

It will be seen that the heavier larger particles will obviously be less influenced by the ambient current and will fall upstream with respect to the collection device. The gold particles and coarser sand and fine gravel will be influenced by the subsurface ambient current and will be moved downstream from the point of discharge and fall upon the collection device.

The ambient current will cause the lighter weight particles to be carried downstream beyond the collection device. Thus, since the current is continuously present, it will be seen that it is merely necessary to adjust the vertical spacing and the downstream spacing between the discharge point and the collection device, depending upon the magnitude of the current.

Test experience indicates that the process permits a high-percentage recovery of the gold particles in the sand and gravel deposits. It is thought that the percentage of gold recovery from the sand and gravel deposits should exceed 92% and is thought to be as high as 95% recovery.

It has also been found that that portion or cut of the sand and gravel deposits that would be directed to the collection device by the action of gravity and the ambient subsurface current is approximately 25% of the entire material constituting a yard of material excavated.

Multi-Wheel Excavator

F. Lachnit; U.S. Patent 3,740,098; June 19, 1973; assigned to Deutsche Babcock & Wilcox AG, Germany designed an underwater excavating device which has a cutting tool to cut into underwater land surfaces, a means for collecting the excavated material, and a provision for conveying the collected material to the collecting device.

A transporting wheel is centrally disposed on the device and has radially outwardly extending edges disposed laterally on each side of the wheel which form a channel around the circumference of the wheel. The conveyor belt, which is adapted to be received by the radially extending edges of the wheel and is disposed on the excavating device so as to be guided in the channel, conveys the collected material from the cutter to the collector.

The advantages of this method of material collection are readily apparent, since the conveyor belt, the transporting wheel, and the radially extending edges form a channel which is almost completely closed for transporting the excavated material. Thus, finely grained material and mud, which are important in scientific testing, are not rinsed away while being transported to the collecting means.

Furthermore, the number of openings between the fixed and movable parts of the device in the range of the material to be collected is greatly diminished. The speed of the conveyor belt preferably corresponds to the circumferential speed of the transporting wheel so that there is a frictionless transportation.

This is achieved by coupling the conveyor belt with the transporting wheel by means of ribs which extend from the edges of the conveyor belt. These ribs engage cut-outs which are provided in the edges of the transporting wheel. During the operation of the device, the ribs of the belt engage the cut-outs and the conveyor belt moves with the same speed as the outer edge of the transporting wheel.

At the same time the ribs avoid the collected material from slipping from the conveyor belt. A further example includes pivotable plates which are mounted on the conveyor belt to scoop up watery mud into the receiving means without any loss of material.

When the device is used for collecting high-capacity material for commercial purposes, and mud and clay deposits are not desired, the conveyor belt is provided with openings which permit the undesired mud and clay deposits to be rinsed away from the conveyor channel during the excavating operation. Thus, only the desired grain material is collected in the receiving means.

When the device is used for collecting materials for scientific purposes, a plurality of collecting drums are usually provided. When the device is used for collecting high-capacity material, the receiver is usually formed like a funnel. Means might also be coupled to the receiving means for continuously conveying the collected material to the water surface.

The device may also be provided with a plurality of collecting drums when it is used for collecting high-capacity material, in order to increase the amount of material to be collected.

Endless Bucket Dredge with Articulated Ladder

C.E. McKay and G.P. Barker; U.S. Patent 3,734,564; May 22, 1973 describe the design for a deep-digging floating dredge which has an articulated ladder with two or more sections pivoted together for relative swinging movement in a vertical plane only and a digging bucket line supported by the ladder. The dredge ladder may have a fixed uppermost section, with either the same or a different bucket line.

For sea-going use the dredge may have sounding means for determining instantaneously the height of the hull above the bottom of the water on which the hull is floating, angle sensing means for determining instantaneously the angle to the horizontal of the upper movable ladder section, and control means receiving an input depth signal and an input angle signal controlling the suspension length of the articulated ladder sections and accommodating it to swells in the water level on which the hull floats.

This design makes deep digging — down to two hundred feet and more below water level — quite practical and economical. The large cross-sectional ladder girder thicknesses and the excessive weights that would be required by previous structures are avoided and made unnecessary for using articulated structures; thus, these ladders are provided in a plurality of sections, two or three usually being sufficient, depending upon the total depth needed.

By combinging these plural sections with adequate rigging lines and maneuvering lines, the depth can be extended at lower cost. This becomes quite significant in locations where it is known that there are extensive deposits of precious metals, gems, commercial minerals, sand, or gravel, or any other subaqueous deposits of commercial value.

Not only can greater depths be attained, but in offshore dredges compensation can be made so that the digging engagement of the dredge buckets, well below the surface, is substantially unaffected by swell and wave conditions.

This does not mean, of course, that dredging should be carried on when the seas become rough, but in other than rough or stormy conditions, this type of ocean-going dredge can operate efficiently even with relatively high swells and even where the minerals lie well below the surface of the sea.

A compensating mechanism which is sensitive to swells and waves and which acts to change the angle of the dredge ladder and its approach sufficiently to keep the digging buckets in engagement with the material being dredged is provided.

Deep Water Dredging Apparatus

In an apparatus designed by *F.A. Kuntz, Jr.; U.S. Patent 3,763,580; October 9, 1973; assigned to Global Marine Inc.*, a surface stratum of an ocean bottom is dredged by a dredging tool connected to the lower end of an elongated hollow conduit pendulously supported from a floating vessel. A plurality of tool guiding lines extend from the vessel to the conduit via respective guides anchored at locations spaced about the region to be dredged.

The effective lengths of the guidelines are adjusted from respective winches mounted on the vessel to sweep the dredging tool back and forth across the region to be dredged. The conduit is rotated about its vertical axis to operate the dredging tool, and an airlift sub injects a stream of air under pressure into the conduit above the dredging tool to lift the cuttings removed from the ocean bottom up through the hollow interior of the conduit.

Briefly, the dredging system includes a vessel floating over a desired region of a surface stratum to be dredged. Preferably, the vessel is a ship of hull form, such as a conventional offshore drilling vessel. When drilling in ocean water, a vessel of hull form provides better stability and seakeeping characteristics than dredging barges which are practical for shallow water, but which are unsafe in ocean water because of their hull form and low freeboard.

The surface stratum is engaged with a dredging tool connected to the lower end of an elongated conduit which is pendulously supported from the vessel. A plurality of tool guiding lines are connected from the vessel to the conduit adjacent the tool via respective locations spaced about the region to be dredged. The effective lengths of the guiding lines are adjusted to locate the tool at a desired position in the region to be dredged by displacing the tool relative to the vessel.

Preferably, the conduit may be a string of oilwell drill pipe coupled to a rotary table mounted on the vessel. The string of drill pipe has lateral flexibility and is therefore capable of covering a relatively large area of surface stratum in deep water even though the vessel remains relatively stationary. However, the conduit is not intended to be limited to a string of drill pipe.

Dredging operations carried out using this apparatus may use a string of relatively thin-walled sections of pipe in place of oilwell drill pipe casing, because dredging operations do not subject the conduit to the high amounts of torque experienced when drilling an oilwell deep into a geologic formation. The lighter weight of a thin-walled pipe also adds to the flexibility of the conduit and makes it substantially easier to guide the dredging tool about the ocean bottom.

In a preferred form, the conduit is rotated about its length to rotate the dredging tool which excavates the surface of the desired underwater region. Preferably, the dredging tool is swept laterally back and forth across the underwater dredging site as the conduit is rotated by cooperatively taking in and paying out the guidelines which are strung from the vessel to opposite sides of the conduit adjacent the dredging tool via respective sheave blocks anchored at spaced-apart locations on the ocean bottom.

The dredging tool is preferably in communication with the hollow interior of the conduit, and during dredging operations cuttings removed from the underwater dredging site are lifted upwardly through the interior of the conduit to the water surface where they are collected. The dredged material is preferably lifted upwardly through the conduit by an airlift mechanism in which a stream of air under pressure is injected into the conduit above the dredging tool to force the dredged material to flow upwardly through the conduit.

STUDYING THE ENVIRONMENTAL IMPACT

EFFECTS OF SURFACE-DISCHARGED DEEP-SEA MINING EFFLUENT

The following information is from PB 226 013. During July and August 1970, a prototype deep-ocean mining system using airlift pumping was tested for mining manganese nodule deposits on the Blake Plateau. Experiments were done, both on board the mining vessel and at the laboratory on shore to determine the environmental effects of surface-discharged deep water used to transport the mined material. A dye-release experiment showed that discharged water mixed slowly with surface water and remained at the surface which was in accord with observed hydrographic data.

Analysis of the raised deep water showed high nutrient levels, which supported high levels of phytoplankton growth and productivity. Sedimentary material discharged with the water enhanced phytoplankton growth. Organic material entrained in the deep water consumed approximately 80% of available oxygen when stored in the dark. In light a spontaneous phytoplankton bloom developed which reversed the oxygen depletion within a few days. No adverse environmental effects could be demonstrated.

RESULTS OF TWO MANGANESE NODULE MINING TESTS

The following information is from PB 218 948. The proposed mining of manganese nodules from the deep-sea floor has triggered a collaboration among industry, the government and academic insititutions to determine the environmental impact of the proposed mining operations before their start. In this way it could be possible greatly to reduce or eliminate completely potential environmental hazards due to the mining operations. This collaboration may lead to the development of mining techniques having such beneficial environmental effects as artificial upwelling.

It is, therefore, of the utmost importance that a very careful study of the environmental impact of projected mining techniques be made, and that adequate baseline information on the physical, chemical and biological environment of potential mining areas be gathered. A general approach to the study of the environmental impact of deep-sea mining; observations made during a suction-dredge mining test held in the North Atlantic Ocean (Blake Plateau) in the summer of 1970; and the study of the environmental impact of a continuous line bucket (CLB) dredge mining test held in the Pacific in August-September, 1972 are covered here.

General Approach

The proposed mining of ferromanganese deposits from the deep sea will have a measurable effect on benthic and pelagic environments within the mining areas. As now envisioned, deep-sea mining operations will result in the removal and redistribution of sediments and benthic organisms and, where suction dredging is employed, in the discharge of nutrient-rich, sediment-laden bottom water into the surface layer or at other depths in the water column.

The research program consists of (1) the establishment of physical, chemical and biological baseline environmental conditions in potential mining areas; (2) the documentation of changes induced in benthic and pelagic ecosystems by deep-sea mining; (3) the elucidation of their underlying mechanisms and implications in relation to current and potential marine resources; (4) the formulation of guidelines for future mining operations which will minimize harmful environmental effects while enhancing the development of potentially beneficial by-products; and (5) the determination of the properties which should be monitored during deep-sea mining to provide the information needed to evaluate the environmental impact of specific mining methods and to devise mitigating measures, if necessary.

The pattern of near-bottom circulation and mixing, and the behavior of "new" water masses formed when bottom water is discharged into overlying water masses will affect local concentrations of organic and inorganic compounds in the water column as well as the distribution and abundance of benthic and pelagic organisms. It will be necessary to determine what changes occur, as well as the time required for the perturbed systems to return to their original state. A time-series study will be undertaken of the following parameters in the mining area:

(1) The character, origin and circulation of the bottom water which is in contact with the manganese pavements.

(2) The volume and dimensions of the transient water masses created by the admixture of bottom water and surface water, and the rate at which these patches lose their identity due to advection and turbulent mixing.

(3) The concentrations of inorganic nutrients, oxygen, trace metals and particulate organic material in the effluent, the alkalinity and the pH of the effluent and receiving waters, their distribution in the discharge area, and the rates at which these parameters are changed in the water mass as a consequence of phytoplankton growth and other biological and physical processes.

(4) The residence time of fine sediments in the surface layer and its effect on light penetration and the concentration of dissolved inorganic nutrients and trace metals, and the residence time of disturbed sediments in the near-bottom layer.

(5)   Changes in the taxonomic composition, distribution and abundance
of benthic and pelagic organisms; the underlying mechanisms of
these changes; and the degree to which they are reversible.

Suction-dredge deep-sea mining will (1) stir up sedimentary material as the
dredge-head sweeps the ocean floor; (2) destroy the benthic organisms and their
habitat in the path of the dredge-head; and (3) introduce sedimentary material,
associated organisms and bottom water into the surface water.

The environmental effect on the water column of surface-discharged sediment-
laden bottom water during a pilot suction-dredge mining test on the Blake
Plateau in the summer of 1970 was estimated.

The temperature and salinity of the water discharged at the surface, or at an
intermediary level in the water column, and the rate of mixing of the discharged
water with the in situ water at the point of discharge, will determine the salinity
and temperature of the resulting mixture and hence its density and stratifica-
tion in the water column. Under the experimental conditions employed during
the Blake Plateau test, the bottom water was discharged at the surface and remained in the euphotic zone, mainly due to some warming in the airlift pumping
process and to the lower salinity of the bottom water.

In manganese-nodule areas in the Pacific, temperature and salinity conditions
of the bottom water will vary. A computer program is now being developed
to predict the density of mixtures resulting from the discharge of bottom water
at the surface, or at intermediate depths, at different rates on a worldwide basis.
Obviously, modifications of the discharge technique could greatly influence the
behavior of the discharged bottom water; it might be sprayed over the sea surface to ensure rapid mixing and warming.

The effect of the discharge of nutrient-rich, sediment-laden bottom water on
oxygen concentration in the surface water was demonstrated to be negligible
in the Blake Plateau experiment.

Results indicated that the development of anoxic conditions was very unlikely
under conditions like those observed on the Blake Plateau because the discharged
water remained at the surface where it can support the growth of oxygen-pro-
ducing phytoplankton; moreover, low levels of oxygen are not likely to be
reached in the open sea because mixing and free exchange of oxygen from the
atmosphere would preclude the formation of an anoxic layer.

Some recent findings made on R/V Conrad Cruise 15-10 in July, 1972, in a
manganese nodule area (Bermuda Rise, North Atlantic) have indicated that several species of small pennate diatoms exist on the sea floor (at depths in excess
of 6,000 meters) which are viable and grow when exposed to light. If this is
a general phenomenon, the discharge of bottom sediments in the photic zone
will seed these waters with algal species not usually found in oceanic phytoplankton communities. This could affect the species composition of phytoplankton in mining areas as well as the food-chains which these organisms
initiate.

Experiments were run during the Blake Plateau test to determine the effects
of discharged bottom water on phytoplankton growth. Bottom water enrich-

ments in excess of 10% were required to stimulate growth. Since the actual concentration of bottom water in the surface layer which resulted from this mining experiment was less than 0.3%, it was concluded that the discharge of bottom water into the photic zone would probably not have a significant effect on phytoplankton growth.

Recent and more detailed experiments lead to the belief, however, that any enrichment, regardless of initial dilution, will enhance phytoplankton productivity. The degree of enhancement will be a function of (1) the nutrient concentration of the discharge; (2) the volume and rate of discharge; (3) the rate at which the effluent is diluted by mixing with surface water, and (4) the species composition and standing crop of phytoplankton in the receiving water and in the discharge.

The overall conclusions from the Blake Plateau experiment were that, under the experimental conditions (1) the discharged bottom water remained in the euphotic zone; (2) it was most unlikely to produce anoxic conditions; and (3) it would significantly increase phytoplankton growth only if the concentration of deep water, after mixing with surface water, was considerably higher than 0.3% in the resulting mixture (0.3% was the maximum concentration of discharged bottom water at the surface during the mining test observed).

Because no research vessel was available to work closely with the mining vessel, the results of this study were incomplete since (1) no benthic study was undertaken, (2) no study of near-bottom currents and turbidity was conducted, and (3) no trace-metal study was conducted.

Therefore, quantitative measurements of the benthic biomass in a manganese nodule province on the Bermuda Rise was undertaken in July, 1972. Conclusions, based on an admittedly very small number of samples, is that the benthic fauna in this area is extremely sparse: average biomass determined in the quantitative samples from this cruise was 9 mg/m^2 of bottom area. The mining would, therefore, affect a small quantity of animals in absolute terms, although its relative impact may be important.

CLB Mining Test of August-September 1972 in the Pacific

The major environmental effects of a CLB mining system would be (1) stirring up sedimentary material as the buckets dredge nodules from the ocean floor, (2) destruction of benthic organisms and their habitat, and (3) introduction of sedimentary material into the entire water column as the buckets are hauled to the surface.

In August and September, 1972, the environmental impact of a CLB mining test in a siliceous ooze province in the North Pacific was observed aboard the R/V Kana Keoki. The physical, chemical and biological conditions of the overlying water column were measured and the benthic fauna at the test site before, during, and after the mining operations was observed.

No evidence of any disturbance in the water column was detected. The dry weight of the particulate material in the water overlying the mining test site varied between 10 and 250 mg/m^3 both before and during mining. Similarly, particulate organic carbon and nitrogen concentrations ranged from 8.3 to 49.8 mg/m^3 and 1.0 to 3.6 mg/m^3, respectively. On one station where the salinity, temperature, depth sensor (STD) accidentally hit the ocean floor the particulate material in the bottom bottles was 167,000 mg/m^3, so it is obvious that when a much larger bucket is dragged along the ocean floor, the turbidity of the surrounding water must be increased by several orders of magnitude over the baseline condition. How long this material remains in suspension and what effect it has on the benthic community must be determined by more sophisticated measuring procedures.

The bottom photographs revealed a considerable diversity in benthic fauna. The terrain varied from 0 to 100% manganese nodule coverage to manganese pavement to rocky outcroppings. Evidence of benthic faunal activity, in particular, mounds, burrows and fecal coils, knots and sprirals was shown. In some cases, there was evidence of weak to moderate bottom currents as shown by erosion of fecal material and orientation of stalked organisms such as sponges and tunicates.

Organisms photographed were sponges, gorgonians, actinarians, bryozoans, crinoids, ophiuroids, echinoids, several species of holothurians (the dominant class photographed), mollusks, cephalopods (squid), pycnogonids, decapods (shrimp), tunicates and rat-tailed fish. Some of the burrowing and mound-building organisms appeared to be present in large colonies. These would obviously be disturbed by a bucket-dredge operation. If the mining were done in strip form, leaving some areas undisturbed, then this would allow the mined areas to be repopulated eventually by these organisms. Trigger weights and other equipment that contact the ocean floor are known to make scars and troughs in the sediment and certainly a large bucket chain would have a greater impact.

The design of the buckets is such that they collect primarily nodules with a minimum amount of sediment. Many of the buckets came to the surface with sediment clinging to the outside, but no sediment was observed in the surface waters as the buckets were hauled up.

Future efforts at monitoring mining tests must include equipment that allows knowledge of position in relationship to the mining vessel and the bucket line (e.g., underwater television).

No environmental effect of the CLB dredge mining test was observed or measured in the North Pacific in September 1972 with the methods described here, except for the possible track of the dredge line observed in one of the bottom photographs made. One problem with these measurements was not knowing whether the equipment at the end of more than 5,000 meters of sea cable was actually over the mined area.

PLAN FOR ENVIRONMENTAL IMPACT EVALUATION STUDY

The mining of Mn nodules will result in the removal and redistribution of sediments and benthic organisms as well as the discharge of nutrient-rich, sediment-laden, bottom water into the surface layer when suction (ALP) dredging is used. Research on the environmental impact of deep-sea mining focuses on the mining-induced changes in the benthic and epipelagic environments and on the quality and quantity of dissolved and particulate matter of the mining effluent. The research program is based on a time-series approach, designed to determine what changes occur and the time required for the perturbed systems to return to their original state. The field program in the mining area will consist of predredging, dredging, and postdredging phases. The predredging phase will provide baseline environmental information on the physical, chemical, and biological processes which characterize the Mn-module province immediately before it is mined.

Observations made during and after the mining operation will then be evaluated on the basis of this reference information. The physical, chemical, and biological programs will be closely coordinated to provide an integrated evaluation of the environmental impact of the mining operation and to project the effects of larger scale operations. The physical, chemical, and biological programs will focus on the following areas, respectively:

(a) the near-bottom circulation patterns, mixing and diffusion, and behavior of bottom water discharged at the surface;

(b) the chemical nature of the water column with respect to dissolved inorganic nutrients, trace metals, dissolved O_2, pH, alkalinity, and particulate organic nitrogen; and

(c) the taxonomic composition, distribution, and abundance of benthic and epipelagic organisms, and the phytoplankton productivity and energy flow through phytoplankton-based food chain.

The study of the engineering properties of the sediments in Mn-nodule areas should result in developing sediment-sampling gear and provide advice to the mining companies on modifications of their mining techniques so as to minimize harmful environmental effects and to enhance the possibility of improvements in the marine environment as by-products of deep-sea mining. The operations are subdivided into three phases.

Phase I, beginning in Fiscal Year 1974, was to concentrate on the establishement of physical, chemical, and biological parameters for a target area situated in the siliceous ooze area of the North Pacific and centered around latitude 10°N, longitude 140°W.

Phase II, beginning in Fiscal Year 1975, will concentrate on the observation of pilot and full-scale mining operations and their environmental impact.

Phase III, beginning in Fiscal Year 1976, will be mainly concerned with the establishment of guidelines for deep-sea mining to avoid harmful environmental effects and to enhance beneficial effects of such mining. At the same time, continuous monitoring of ongoing mining operations will begin.

The prospect of extensive deep-sea mining requires serious consideration of the environmental impact of these activities. Such mining should have definite effects on the benthic, and possibly pelagic, environments within the mining area, perhaps significantly altering benthic or pelagic populations and ecological systems. It is essential, therefore, that the environmental impact of Mn-nodule mining be well defined and its implications understood before deep-sea mining is attempted on a large scale.

LDGO has been actively involved with these environmental impact studies; to date, several preliminary in-field investigations have been conducted and have developed a comprehensive plan for an eventual, complete environmental evaluation of deep-sea Mn-nodule mining. The overall objectives of our environmental impact study of deep-sea mining are:

(a)   the establishment of physical, chemical, and biological baseline-environmental conditions in potential mining areas;

(b)   the documentation of changes induced in benthic and pelagic ecosystems by deep-sea mining;

(c)   the elucidation of the underlying mechanisms and implications in relation to current and potential marine resources;

(d)   the formulation of guidelines for future mining operations which will minimize harmful environmental effects while enhancing the development of potentially beneficial by-products; and

(e)   the determination of the properties which should be monitored during deep-sea mining to provide the information needed to evaluate the environmental impact of specific mining methods and to devise mitigating measures, if necessary.

The mining of nodules from the ocean basins involves potential changes in the ocean-floor environment and its associated benthic organisms as well as within the environment of the entire water column separating the ship from benthos. Several different methods are contemplated for the retrieval of Mn nodules; and, although some common effects can be expected from all of them each method should present some unique potential problems of its own. In general, we should expect all mining techniques to:

(a)   destroy the benthic organisms and their habitats in the path of the mining operation;

(b)   stir-up sedimentary material as the mining implement sweeps the ocean floor; and

(c)   introduce sedimentary material, associated bottom organisms, and bottom water into various layers of the water column, including, in some cases, the surface water.

The degree to which each of these factors occurs is dependent on the mining techniques employed. The CLB system will introduce little sediment or bottom material into the upper portion of water column (especially at the surface), while the ALP system of dredging will specifically discard sediment and bottom water into the surface waters. Likewise, hydraulic mining should introduce much more sediment into the lower water column than that of suction dredging.

The destruction of benthic organisms and their habitats is definitely a problem; the extent of this destruction and the capacity for benthic organisms to reestablish themselves within a reasonable length of time will determine how serious the effect will be. The stirring-up of bottom sediments could also clog or smother benthic organisms over a much wider area than that which is actually mined. If this were a significant problem, it would make repopulation by the sessile benthic organism even more difficult and would make ineffective any techinique of leaving areas or strips unmined for repopulation purposes, unless the areas left intact were very extensive. The introduction of bottom water and material into the upper water column is more complex and may prove either beneficial or harmful.

Introduction of sediments and bottom material into the surface waters may increase trace-metal concentrations which could inhibit photosynthesis or allow
the accumulation of different trace metals within marine food chains. More
likely, however, an increase in photosynthetic activity and productivity will result from the high nutrient concentration of the bottom water.

The concentration of nutrients that is introduced into the surface water and the
amount of time that this nutrient-rich water mass remains in the euphotic zone
will determine the extent of its effect. A phytoplankton bloom should be beneficial if a food chain, including larger marine organisms, also develops; if not,
fouling by decaying plankton may result.

In the summer of 1970, the environmental effect of the water column of surface-discharged sediment-laden bottom water was estimated during a pilot ALP
mining test on the Blake Plateau. In this test, the fate of the bottom water discharged into the surface waters was studied as well as effects of such water on
the in situ O_2 concentration and phytoplankton growth.

Aerial photographs of the dye-labeled discharge water showed the deep water remaining in the upper 10 meters of the water column 3 hours after the discharge.
The density of the discharged deep water, relative to the in situ waters, determines
whether this water sinks. Density is dependent on the salinity and temperature
of the water mass, and a computer program should be able to predict the density of mixtures resulting from the discharge of bottom water at various depths
in the water column.

Tests on the O_2 content of the deep water during dark and light incubation in
carboys suggested that it was most unlikely for such discharged water to produce
anoxic conditions. Also, the effect of the bottom water on phytoplankton
growth suggested that significant increases in growth would only occur if the
concentration of the deep water, after mixing with surface waters, was considerably higher than the 0.3% which was found to result from the discharge in the
test that we monitored.

In July 1972, a cruise aboard the R/V Robert D. Conrad was undertaken to
determine physical, chemical, and biological baseline conditions in a Mn-nodule
province on the Bermuda Rise. The benthic biomass, deep-water nutrients,
physical characteristics, and phytoplankton growth in surface water as a function
of enrichment with bottom water was studied.

The benthic fauna was found to be very sparse in this area, averaging only
~9 mg/m^2 of bottom area. Mining in such an area would, therefore, affect a
small quantity of animals in absolute terms, although the relative impact of the
mining may be important. The enrichment studies showed phytoplankton growth
to be directly dependent on the degree of bottom-water enrichment. In addition
bottom water and sediments incubated without the addition of surface water
showed a phytoplankton bloom which suggests the possibility of spores in the
bottom sediments introducing new species of phytoplankton into the surface
waters.

In August and September 1972, a test of the CLB mining system in a siliceous
ooze province in the North Pacific was monitored; monitoring of the Japanese
mining vessel Kyokuyo Maru 2 was accomplished from the University of Hawaii's

R/V Kana Keoki. The physical, chemical, and biological conditions of the overlying water column was measured, and the benthic fauna at the test site was observed and sampled before, during, and after the mining operation. Temperature,
salinity, O_2, turbidity, and nutrients (NO_3, NO_2, SiO_4, NH_4, and PO_4) as well as the
dry weight, C, and N content of the particulate matter in the water were meas-
ured to within 10 meters of the bottom.

In addition, three cores of the bottom were taken, and several camera stations
provided many bottom photographs. Camera station observations and data on
the water column were obtained both before and during mining operations. In
this preliminary work, no definite effects of mining have been observed, ten-
tatively suggesting that mining disturbances were not great.

A technical plan for a complete evaluation of suction-dredge (such as ALP)
mining is presented here. An environmental impact study of suction-dredge
mining is the more complex because extensive monitoring of effluent effects on
the upper water column is also necessary. This plan will, therefore, define all
the techniques, instruments, and equipment which are required to accomplish
a thorough and complete environmental impact study.

One or two 1-month cruises for Fiscal Year 1974 were proposed to: (a) establish
baseline conditions for the biological, chemical, and physical parameters; (b) in-
vestigate more thoroughly the methods of sampling; and (c) characterize and
become familiar with this area of the Pacific. One month's ship time would
allow 10 days for travel to and from the station and approximately 2 weeks for
the baseline characterization. The target area chosen for this work, the siliceous
ooze of the North Central Pacific, was chosen because:

(a) a good cover of Mn nodules exists;

(b) the conditions in general are favorable for mining;

(c) some preliminary observations of Mn-nodule mining in this general
     area have been done; and

(d) the currents tend to make this a very placid area and, therefore, a
     good test site for dispersal in calm waters.

The baseline study concerns itself basically with three general areas; observation of the benthos; observation of the lower water column (within ~500 m
of the bottom); and observation of the upper portion of the water column (the
euphotic zone, surface to ~500 m). Some measurements and samples, however,
will be routinely taken over the whole length of the water column; these include
current studies, STD, and turbidity profiles.

PROCEDURE FOR SAFE DEEP-SEA MINING

The following information is taken from the report presented by O.W. Roels at
the Tenth Annual Conference of the Marine Technology Society held in
Washington, D.C., September 23-25, 1974. The proposed mining of manganese
nodules form the deep-sea floor has triggered a perhaps unique collaboration in
the United States between the government, mining industries, and academic institutions to determine the environmental impact of such mining operations before their start. By taking preventive action, it should be possible to greatly re-

duce or completely eliminate potential environmental hazards due to the mining operations.

The areas of greatest commercial interest are situated in the north equatorial central Pacific in water depths between 3,000 and 6,000 meters, at least 1,500 kilometers from shore. Manganese nodules are generally found in areas of extremely slow sedimentation of red clay and siliceous (radiolarian) ooze. The nodules lie mainly on top of the sediments covering the ocean bottom underlying oceanic water masses of very low biological productivity; therefore, no deep penetration of the sediments will be required to retrieve them. Manganese nodules are rare in areas where there is rapid sedimentation, e.g., on those parts of the sea floor underlying areas of high biological productivity in the water column, giving rise to rapid sedimentation of biogenic oozes.

The areas to be mined will be limited, therefore, by the distribution of manganese nodules on the ocean floor and by technical and economic factors governing their retrieval from the depths. Therefore, only relatively flat, sediment-covered parts of the ocean floor with a high density of manganese nodules on, or very close to, the surface of the sediment are being considered.

Nodule Collection Systems

In the first steps of the mining operation, the manganese nodules are collected from the ocean floor, usually from great depths, and transported through the water column to a surface vessel.

A variety of systems propose the use of bottom-gathering devices connected to hydraulic or airlift pumping systems for transporting the nodules to the surface through a pipeline system. All of these machines have components which contact the ocean bottom to make a first separation of the nodules from the surrounding sediment. This first separation is achieved by a chute with water jets, heavy spring-rake tines, a radial tooth roller, harrow blades and water jets, or spaced comb teeth.

Many devices employ adjustable collecting elements so that changes can be made during the mining operation to accommodate variations in the nodule-deposit and sediment characteristics. The continuous-line-bucket (CLB) dredge system, tested in the Pacific in 1971 and 1972, used buckets of 40 cm depth with a maximum penetration into the sediment of about 20 cm, probably much less in practice. All of the different techniques under consideration will attempt to avoid, as much as possible, the retrieval of sediments with the nodules. An important feature of all of the collecting devices is a controlled digging depth into the ocean bottom: interest is usually centered within the upper few inches of the sediment.

Seafloor Disturbance and Sediment Redistribution

It is in the interests of a mining operation to attempt separation of nodules from sediment on the ocean floor, and to disturb the sediment as little as possible where compatible with efficient collection of the nodules. There will obviously be significant disturbance of the sediment, resuspension of sediments in near-bottom waters, and destruction of sessile benthic organisms which cannot escape the oncoming dredge. A clout of sediment will undoubtedly be stirred up in the near-bottom water layers. The distribution and resedimentation of the stirred-up particles will be governed by their density and other sedimentation characteristics, as well as by the near-bottom currents. This resuspension of sedimentary materials will affect the near-bottom water mass, certain areas of the ocean floor from which sediments have been removed, as well as other areas where redeposition will occur.

The near-bottom water mass may retain in solution certain compounds leached out from the sediment or from the interstitial water. For instance, it is conceivable that the trace-metal content of the near-bottom water could be increased by this stirring up of the bottom and resuspension of sediment. The enrichment in certain compounds of the near-bottom water may have a stimulatory or an inhibitory effect on organisms living near the sea floor in the deep ocean.

It has been argued that the redistribution of sediment on the ocean floor resulting from natural phenomena exceeds by many orders of magnitude, on a worldwide scale, any disturbance by all of the dredges ever likely to be utilized in deep-sea mining. Overall, severe effects of manganese-nodule mining seem unlikely, in view of both the apparently low density of near-bottom prowlers and the fact that the sedimentary material arrived on the seafloor as a result of natural sedimentation processes.

However, it remains equally clear that local disturbance of sediment will have a certain impact on the deep-sea fauna and flora. This is particularly the case for sessile animals which may have a very slow reproductive cycle. It is unlikely that any mining operation will (nor should they) cover 100% of a given area of the seafloor: thus, seafloor bands of adequate width should be left undisturbed in a mined area to enable the reestablishment of deep-sea fauna and flora in those areas where the dredge heads have destroyed it. This process of recolonization would be quite rapid on a geological timescale.

It is believed that the biomass of sessile fauna on the deep-sea floor is generally small, particularly in manganese nodule areas and, therefore, the quantitative impact of deep-sea mining on marine flora and fauna should be quite small.

Another possible result of the disturbance of the sediments and their resuspension in the water column, is the transplantation of spores or other dormant or live forms of microorganisms from one area, where they rest in the sediment, to another, transported by water currents in the overlying water masses after resuspension from the dredged sediments.

Surface Discharge of Sediment and Near-Bottom Water

After the manganese nodules have been collected from the seafloor (together with certain quantities of sedimentary material) they are transported through the water column to the surface mining vessel, either in the buckets of a CLB system or in a water stream through a pipeline (e.g., airlift). In both modes of transport, some or all of the accidentally gathered sediment and near-bottom water may be discharged, either at the surface or at intermediate depths in the water column. The effect of these discharges at the surface has been measured or forecast. To date, however, there is no information concerning the rate of sedimentation of such discharged particulate matter, although there is some information available concerning the influence of deep-sea sediments on the productivity of waters in the euphotic zone. The influence of dissolved nutrients from interstitial water, or from near-bottom water, on the chemical composition of the overlying water column can be calculated from the rate of mixing and the rate of near-bottom water discharged at the surface. Mixing will not only be governed by the salinity and temperature of the near-bottom water at the time of discharge, but also by the salinity and temperature of the receiving water mass.

Consideration should also be given to the possibility of introducing foreign species of phytoplankton to the surface and upper water column, species which were dormant in the sediments but which may revive when discharged into suitable temperature, light and oxygen conditions.

From the admittedly incomplete results of the work done so far, it appears that the effect of the mining operations and of the vertical transport of nodules, sediment and near-bottom water to the surface and its discharge at the surface or at intermediate levels in the water column is small.

At-Sea Processing Wastes

The at-sea processing and extractive metallurgy of manganese nodules and the discharge of waste materials resulting from such processing, could be far more dangerous to the marine environment, unless adequate precautions are taken. Most major concerns involved in the development of manganese-nodule mining have determined that, at least for first-generation plants, economical processing can only be accomplished ashore.

The principal reasons for this are that the reagent transportation costs will be equal to, or greater than, the nodule transport costs, and problem of waste disposal and environmental protection will be much greater at sea than on land. However, if all processing takes place at sea, the care taken in waste disposal resulting from metallurgical processes should be, at the very least, equal to that of land-based operations of similar nature. To ensure the safe development of this marine resource, the following procedure might be adopted:

(1) The establishment of baseline conditions in the potential mining areas.

(2) The environmental monitoring of pilot and/or fullscale mining operations.

(3) The documentation of changes induced in benthic and pelagic ecosystems by deep-sea mining and evaluation of their implications in relation to current and potential marine resources.

(4) If necessary, the recommendation of changes in mining methods and equipment use, based on the facts established in (2) and (3).

(5) The formulation of environmental criteria and regulations for future mining operations to minimize harmful environmental effects while enhancing the development of potentially beneficial by-products.

(6) The monitoring and enforcement of (5).

This procedure could be implemented rapidly by most interested nations and should then serve as a model for possible international adoption.

ECONOMIC CONSIDERATIONS

ASPECTS OF DEEP OCEAN MINERALS EXPLOITATION

The following information is taken from Report AD 734 968.

Prospective World Aggregate Mineral Demand

In 1968, the world population was estimated to be 3,509,100,000. Over 2.5 billion or about 72.5% of this population resided in developing countries, defined as: all the states in Latin America, East Asia and the Pacific less Japan and Australia, South Asia and the Indian Ocean, the Near East, Africa less the Republic of South Africa, Albania, Bulgaria, Greece, Portugal, Spain, Turkey and Yugoslavia. Population projections for the year 2000 are 6,389,000,000 for the world, with 4,777,000,000 or about 74.8% belonging to the less developed countries.

By the year 2020, the world's population is expected to increase to 9,025,000,000 while the share of developing countries will be an even greater percentage. On the basis of these population statistics alone, one could expect the world demand for minerals to increase 174% by the year 2000 and 257% by 2020. It is also reasonable to assume that changes in the values and tastes of the world's people will require higher standards of living, causing greater demands for raw materials.

Predictions of the world aggregate of gross national product (GNP) are another indication of future demands. In 1968, the total GNP for the world was $2,685,006,000,000. Developing countries accounted for $450,066,000,000 of this or only 16.8% of the world total. Gross national product predictions for the world of 2000 total $10,848,000,000,000 with the developing nations producing $1,589,000,000,000 or 14.6% of the total.

By the year 2020 the world's GNP figures are expected to reach the neighborhood of $28,711,000,000,000. With these figures, future per capita GNP can be estimated and used as an indication of mineral demands.

TABLE 5.1: PER CAPITA GNP PREDICTION

	1968	2000	% Increase 1968 - 2000	2020	% Increase 1968 - 2020
World total	$ 865	$1,690	195	$3,182	368
Developed nations	2,312	5,770	249		
Developing nations	177	321	182		

If future mineral demands were assumed to increase proportionally to the world's GNP, they would be 407% greater than the 1968 demand by the year 2000 and 1,069% higher in 2020. A substantial reduction in the above projections, however, must be calculated because of the disparity in per capita GNP between people in developed and developing nations and the deescalating rate of industrialization in developed countries owing to a larger portion of income being directed towards education, services and leisure.

Other considerations, such as the substitution for current mineral uses by synthetics, increased dependence on recycling scrap materials as a source of mineral supply, and the possibilities of war or famine would greatly affect future demands for raw materials.

These conditions are impossible to predict with any precision. Ignoring them and assuming that demand will increase at about 75% of the GNP rate, a reasonable projection for the year 2000 would be 300% of 1968's demand and by 2020 some 700% more than 1968.

Prospective Mineral Reserves

Phosphorite: Phosphorite nodules have been dredged at depths up to 11,400 feet at the base of the continental slope. However, the nodules were probably carried to these depths by slumping or turbidity currents from larger deposits resting under shallower water on the continental shelves. Since any propensity for mining these nodules would be directed toward the continental shelves, phosphorite deposits can be eliminated as prospective deep ocean mineral reserves.

Calcareous Oozes: Calcareous ooze, particularly globogerina ooze, has been mentioned as an alternative source of cement rock because it has the required chemical content and necessary physical characteristics; fine grained and unconsolidated with a large surface area.

Globogerina ooze covers approximately 35% of the ocean floor with an average thickness of 400 meters. It is a substantial resource, but cement rock in various forms is plentiful on land except for shortages in certain local areas. In any event, the low commercial value of calcareous ooze precludes its economic exploitation from the deep ocean.

Diatom Ooze: Diatom ooze, covering almost 9% of the ocean floor with an average thickness of 200 meters, could be used as a substitute for diatomite. The assay of diatom ooze is favorable for this purpose as are its physical properties. It is not likely, however, that diatom ooze will become economically competitive with diatomite, because the current value of diatomite is less than $3.00 per ton at the mine site. Diatomite also has other substitutes, such as perlite and vermiculite, which are already replacing it for some uses.

Red Clay Deposits: Red clay, covering about 28% of the ocean floor with an average thickness of 200 meters, is basically composed of hydrated aluminum silicates. Assays of up to 25% aluminum oxide make it comparable to some land deposits of aluminum ores. However, land deposits are extensive and generally more valuable. Using red clay as a construction material is also a possibility, but available terrestrial clay deposits are far more economic.

Red clay is usually contaminated with an average of one to two percent manganese grains. These grains are also associated with minute quantities of copper, nickel and cobalt. Because of the low concentration of these grains in red clay, their exploitation would not be economically desirable.

Red Sea Metaliferrous Muds: Heavy metal deposits were recently found in the Red Sea in conjunction with the mid-ocean ridge system extending into that sea. Deposit depths are from 2,000 to 2,170 meters and cover a total area of approximately 75 square kilometers.

Exploration to date is not extensive. Only eighty cores have been taken in the deposit region to a maximum depth of 10 meters. Six facies have been identified with varying metal assays. A metaliferrous sulfide facie, penetrated by only four cores thus far, appears to have the richest assay.

A considerable amount of new technology is needed before exploitation of these metaliferrous muds becomes feasible. Mining technologies would have to be achieved for selective dredging of the best facies, dewatering the extracted muds and disposing of waste material. Even when mined material can be taken to a processing site, new techniques must be found to extract the metals from this unusual ore.

As a result of intensive exploration efforts and continued technological research, the Red Sea metaliferrous muds will undoubtedly become a new reserve for metals of world-wide importance. The economic value of these deposits cannot be determined with any accuracy until mining and processing techniques are further researched and the costs and benefits evaluated with more certainty. Progress will take time. Although mining rights have been applied for by several groups, and others have indicated an interest, production should not be expected in the near future.

Mid-Ocean Ridges: Since the conditions under which the Red Sea metal deposits were formed are associated with the mid-ocean ridge system, it is reasonable to assume that similar conditions may exist along the approximately 40,000 miles of mid-ocean ridges throughout the world. As yet, no similar deposits have been located, but the possibility of their existence cannot be ruled out until an extensive exploration of the mid-ocean ridge system is undertaken.

Additionally, there are potentialities for copper, concentrated by metamorphism, to exist along the mid-ocean ridge system, while similar deposits of chromite or platinum may be found in peridotite. Such possibilities are speculative; no evidence confirms their existence.

Manganese Nodules: The presence of manganese nodules on the deep ocean floor is known since the Challenger Expedition of 1873-1876. The nodules vary in size from small grains to one of 1,770 pounds recovered east of the Philippine Islands. They can be found almost anywhere in the deep oceans. Tonnage estimates for the Pacific Ocean alone reach 1.66 trillion tons. Nodule density varies over the ocean bottom.

Reported ranges on the Pacific Ocean floor are from less than 0.1 pound per square foot to over 7 pounds per square foot, and standard deviations of 1 pound per square foot are common within a locality. Concentrations of nickel, manganese, copper, cobalt and possibly other metals having commercial exploitability are locked into complicated hydrated oxides within the nodules. While the existence of this treasure has been known since the late 1800's, the feasibility of its exploitation has only recently been realized.

Metal assays for manganese nodules vary from location to location, making deposits in certain areas more favorable than others. In order to determine thoroughly the distribution of nodule types, extensive sampling will be required. With the recent efforts to exploit deposits of manganese nodules, faster and less expensive sampling techniques are being developed.

Better knowledge of nodule formation will also help in predicting nodule grades, as variations are believed to result from differences in initial formation processes, subsequent physical and chemical erosion, seawater element concentrations, hydrostatic pressures, temperatures and, possibly, bacteriological activity.

With respect to economic exploitability, nodules in the Pacific Ocean are generally of higher grade than those of the Atlantic or Indian Oceans. Atlantic nodules contain less manganese, copper, cobalt, molybdenum and titanium, but are usually less variable in composition than Pacific nodules. In the Indian Ocean, the nodules show smaller average assays of manganese, nickel, copper and cobalt, the four primary exploitable constituents.

Exploitation of manganese nodules will require identification of mine sites that provide the best commercial nodules in adequate abundance. The proximity of a mine site to prospective markets will also be a consideration in any site's selection. Each specific mine site will require intensive quantitative and qualitative sampling, exploration of bottom characteristics and searches for bottom obstructions. Currents, salinities, temperatures, sea conditions and other meteorological or oceanographic information must also be known.

Numerous mining techniques have been proposed and at least two have been successfully tested. One of these, a hydraulic suction dredging system, was tested in 2,500 feet of water on the Blake Plateau off the coast of Florida and Georgia in July, 1970. The other system, drastically different, was tested in September, 1970, near Tahiti in 3,760 meters of water from a Japanese ship; it consisted of a mechanical continuous line bucket dredge.

Once raised to the sea surface, transportation of mined manganese nodules to shore can be accomplished by the mining vessel itself, by ore carriers onto which the ore is transferred at sea, or by ocean-going tugs pulling barges. Transport via the mining vessel would require the smallest capital investment, but would cause a loss in mining time and involve raising or disconnecting the mining rig each time the vessel is filled to capacity. Several methods have been proposed for processing nodules, once they have been taken ashore. The constituent metals are in the form of oxides instead of sulfides, as normally found in terrestrial ores. This circumstance requires a completely new chemical engineering approach at refining plants.

Deepsea Ventures, one of the companies doing research in this area, has a pilot plant in operation which is capable of processing one ton of nodules per day. Tentative plans are to enlarge this model until a plant capacity of 3,000 tons-per-day will be reached by 1975 or 1976. This plant will eventually be capable of processing the 1,000,000 tons-per-year that Deepsea Ventures envisages mining.

Effects of Manganese Nodule Mining

Concern has been expressed over the effect large scale mining of manganese nodules will have on the prices of constituent metals. Because of the disparity between the ratio of constituent metals in the nodules and the ratio of their world demands, this concern is warranted. With many different grades of nodules available, the extent of the disparity will be dependent on the type of nodule mined.

The Pacific Ocean, having the best overall commercial grades of nodules, can be divided into areas of high manganese, high nickel and copper, and high cobalt content, Table 5.2 shows constituent metal contents for average-assays of these nodule types, plus the composition of nodules from the proposed Deepsea Ventures' mine site. Values of metal yield per ton of nodules at 1971 prices are also shown in Table 5.2, using a 98% processing yield figure claimed by Deepsea Ventures.

It can be seen that nodules with high copper and nickel or high cobalt content are the most lucrative commercially at current market prices. High manganese content nodules, on the other hand, are typically poor in the other metals and have the lowest commercial value of the types shown.

Metal yields for various grades of nodule deposits are shown in Tables 5.3 to 5.6, using increments of 1,000,000 tons of mined nodules per year as proposed by Deepsea Ventures and assuming a 98% processing yield. These projected figures estimate an approximate 300% increase in world mineral demand by the year 2000.

In order to determine the effect that extensive exploitation of manganese nodules may have on future prices of the constituent metals, one must compare expected metal yields with actual world production. Tables 5.7 to 5.10 show anticipated yields from increments of 1,000,000 tons of mined nodules as a percent of reported 1969 world metal production.

It is obvious from this comparison that the mining of 1,000,000 tons per year commencing in the near future would not overly flood the world market with manganese, nickel or copper. A single operator of such size could, however, have a detrimental effect on the cobalt market, especially if a deposit of high cobalt content were mined and 52% of the world demand for cobalt was produced. This could reduce the current price of cobalt by as much as one third,

TABLE 5.2: METAL CONTENT AND VALUE OF NODULE TYPES

Metal	1971 Market Price	High Mn Content Nodules	High Ni and Cu Content Nodules	High Co Content Nodules	Deepsea Ventures' Mine Site Nodules
Manganese Ore*	$ .03 per lb.	49.80%	33.30%	28.50%	26.00%
Nickel	1.28 per lb.	0.26	1.52	0.66	1.26
Copper	.52 per lb.	0.14	1.13	0.21	1.00
Cobalt	2.20 per lb.	0.055	0.39	1.20	0.24
Value of Metal Yield Per Ton of Nodules at 1971 Prices		$39.72	$86.12	$87.18	$67.43

*Based on 46% Mn Ore.

or from $2.20 to $1.49 per pound. Three such mining operations by 1980 would not only flood the world market with cobalt, but begin an inundation of manganese equal to 9 to 18% (depending on nodule grade) of the world's manganese production in 1969. A yield of cobalt equal to 158% of world demand would clearly be disastrous and such would be possible from three million tons of nodules with high cobalt content.

The mining of manganese nodules at the rate of three million tons per year would also begin to be felt by the nickel market, especially if high nickel content nodules were processed. The copper market appears to be safe from detrimental effect.

A nodule production of over 3,000,000 tons per year probably won't occur before 1980. However, by 1985 the world nodule mining capacity could possibly reach as high as 10,000,000 tons. A comparison of possible mining yields or capacity with projected high and low demands for metal by 1985 is given in Tables 5.11 to 5.14.

Cobalt clearly will suffer the most. Five million tons of nodules with high cobalt content would meet over 200% of the 1985 demand, and ten million tons would push supply to 458% of the world need, causing the value shown in Table 5.2 of cobalt, and the high bearing nodules, to plunge downward. This grade of nodule would then become economically less desirable to exploit, making the high nickel and copper content nodules the optimum type to mine.

The Deepsea Ventures' mine site nodules may be relatively valuable. Because their manganese, nickel, and cobalt assays are generally less than the high nickel and copper content nodules, increasing production rates would not have as great a depressing effect on the price of those constituent metals.

Of course, if the capacities of the terrestrial mining industry do not increase as fast as world demand, exploitation of manganese nodules will not have as great an effect on metal prices. It will serve only to satisfy a portion of the world's greater needs. Such could indeed be the case for all the constituent metals, except cobalt, by 1985. Copper from manganese nodules will certainly fall into this category, as a ten million ton mining capacity could only satisfy a maximum of 1% of the 1985 demand.

Anticipated increases in demand of about 50% from 1985 to the year 2000 render mining capacities of 15,000,000 tons nearly the same in percentile figures as a 10,000,000 ton capacity in 1985. Mining capacities of twenty million and twenty-five million tons give proportionately higher projections as can be seen in Tables 5.15 to 5.18.

It is evident from the above that the future world market for cobalt will be the only metal industry seriously affected by the exploitation of manganese nodules. The other three metals, manganese, nickel and copper, found in nodules will not have their world markets so much affected, but individual mines, localities or countries may feel stronger consequences. The copper market should not suffer any harm whatsoever, while sectors of the manganese and nickel markets will probably be affected to some extent, depending on the quantity and grade of nodules mined and the future terrestrial mining capacities for these two metals.

TABLE 5.3: METAL YIELD FOR HIGH MANGANESE CONTENT NODULES (Tons-Per-Year)

Mining Capacity	Manganese	Nickel	Copper	Cobalt
1,000,000	488,000	2,550	1,370	540
2,000,000	976,000	5,100	2,740	1,080
3,000,000	1,464,000	7,650	4,110	1,620
4,000,000	1,952,000	10,200	5,480	2,160
5,000,000	2,440,000	12,750	6,850	2,700
6,000,000	2,928,000	15,300	8,220	3,240
7,000,000	3,416,000	17,850	9,590	3,870
8,000,000	3,904,000	20,400	10,960	4,320
9,000,000	4,392,000	22,950	12,330	4,860
10,000,000	4,880,000	25,500	13,700	5,400
15,000,000	7,320,000	38,250	20,550	8,100
20,000,000	9,760,000	51,000	27,400	10,800
25,000,000	12,200,000	63,750	34,250	13,500

TABLE 5.4: METAL YIELD FROM HIGH NICKEL AND COPPER CONTENT NODULES (Tons-Per-Year)

Mining Capacity	Manganese	Nickel	Copper	Cobalt
1,000,000	326,300	14,900	11,080	3,820
2,000,000	652,600	29,800	22,160	7,640
3,000,000	978,900	44,700	33,240	11,460
4,000,000	1,305,200	59,600	44,320	15,280
5,000,000	1,631,500	74,500	55,400	19,100
6,000,000	1,957,800	89,400	66,480	22,920
7,000,000	2,283,100	104,300	77,560	26,740
8,000,000	2,610,400	119,200	88,640	30,560
9,000,000	2,936,700	134,100	99,720	34,380
10,000,000	3,263,000	149,000	110,800	38,200
15,000,000	4,894,500	223,500	166,200	57,300
20,000,000	6,526,000	298,000	221,600	76,400
25,000,000	8,157,500	372,500	277,000	95,500

TABLE 5.5: METAL YIELD FROM HIGH COBALT CONTENT NODULES (Tons-Per-Year)

Mining Capacity	Manganese	Nickel	Copper	Cobalt
1,000,000	279,200	6,470	2,580	11,760
2,000,000	558,400	12,940	5,160	23,520
3,000,000	837,600	19,410	7,740	35,280
4,000,000	1,116,800	25,880	10,320	47,040
5,000,000	1,396,000	32,350	12,900	58,800
6,000,000	1,675,200	38,820	15,480	70,560
7,000,000	1,954,400	45,290	18,060	82,320
8,000,000	2,233,600	51,760	20,640	94.080
9,000,000	2,512,800	58,230	23,220	105,840
10,000,000	2,792,000	64,700	25,800	117,600
15,000,000	4,188,000	94,050	38,700	186,400
20,000,000	5,584,000	129,400	51,600	235,200
25,000,000	6,980,000	161,750	64,500	294,000

TABLE 5.6: METAL YIELD FROM DEEPSEA VENTURES' CANDIDATE MINE SITE (Tons-Per-Year)

Mining Capacity	Manganese	Nickel	Copper	Cobalt
1,000,000	254,500	12,340	9,800	2,350
2,000,000	509,000	24,680	19,600	4,710
3,000,000	764,000	37,020	29,400	7,060
4,000,000	1,019,000	49,360	39,200	9,400
5,000,000	1,273,000	61,700	49,000	11,760
6,000,000	1,528,000	74,040	58,800	14,110
7,000,000	1,783,000	86,380	68,600	16,470
8,000,000	2,038,000	98,720	78,400	18,820
9,000,000	2,292,000	111,060	88,200	21,170
10,000,000	2,545,000	123,400	98,000	23,520
15,000,000	3,819,000	185,200	147,000	35,280
20,000,000	5,090,000	246,800	196,000	47,050
25,000,000	6,365,000	308,500	245,000	58,800

TABLE 5.7: METAL YIELD FROM HIGH MANGANESE CONTENT NODULES AS A PERCENT OF 1969 PRODUCTION

Mining Capacity	Manganese	Nickel	Copper	Cobalt
1,000,000	6.0	0.5	0.02	2.4
2,000,000	12.0	1.0	0.04	4.8
3,000,000	18.0	1.4	0.06	7.3
4,000,000	24.0	1.9	0.08	9.7
5,000,000	29.9	2.4	0.10	12.1
6,000,000	35.9	2.9	0.13	14.5
7,000,000	41.8	3.4	0.15	17.3
8,000,000	47.8	3.8	0.17	19.4
9,000,000	53.8	4.3	0.19	21.8
10,000,000	59.8	4.8	0.21	24.2
15,000,000	89.8	7.2	0.31	36.3
20,000,000	119.7	9.6	0.41	48.4
25,000,000	149.6	12.0	0.52	60.6

TABLE 5.8: METAL YIELD FROM HIGH NICKEL AND COPPER CONTENT NODULES AS A PERCENT OF 1969 PRODUCTION

Mining Capacity	Manganese	Nickel	Copper	Cobalt
1,000,000	4.0	2.8	0.2	17.1
2,000,000	8.0	5.6	0.3	34.2
3,000,000	12.0	8.4	0.5	51.4
4,000,000	16.0	11.2	0.7	68.5
5,000,000	20.0	14.0	0.8	85.7
6,000,000	24.0	16.8	1.0	102.8
7,000,000	28.0	19.7	1.2	119.9
8,000,000	32.0	22.5	1.3	137.0
9,000,000	36.0	25.3	1.5	154.1
10,000,000	40.0	28.1	1.7	171.2
15,000,000	60.0	42.1	2.5	256.9
20,000,000	80.0	56.2	3.3	342.5
25,000,000	100.0	70.2	4.2	428.0

TABLE 5.9: METAL YIELD FROM HIGH COBALT CONTENT NODULES AS A PERCENT OF 1969 PRODUCTION

Mining Capacity	Manganese	Nickel	Copper	Cobalt
1,000,000	3.4	1.2	0.03	52.7
2,000,000	6.8	2.4	0.08	105.4
3,000,000	10.3	3.7	0.12	158.1
4,000,000	13.7	4.9	0.16	211.0
5,000,000	17.1	6.1	0.20	263.8
6,000,000	20.5	7.3	0.23	316.3
7,000,000	24.0	8.5	0.27	369.2
8,000,000	27.4	9.8	0.31	421.8
9,000,000	30.8	11.0	0.34	475.0
10,000,000	34.2	12.2	0.39	527.3
15,000,000	51.4	17.7	0.59	791.4
20,000,000	68.5	24.4	0.78	1053.8
25,000,000	85.6	30.1	0.98	1318.5

Ocean Floor Mining

TABLE 5.10: METAL YIELD FROM DEEPSEA VENTURES' MINE SITE AS A PERCENT OF 1969 PRODUCTION

Mining Capacity	Manganese	Nickel	Copper	Cobalt
1,000,000	3.1	2.3	0.1	10.5
2,000,000	6.2	4.6	0.3	21.1
3,000,000	9.4	6.9	0.4	31.6
4,000,000	12.5	9.3	0.6	42.2
5,000,000	15.6	11.5	0.7	52.7
6,000,000	18.7	13.9	0.9	63.3
7,000,000	21.8	16.1	1.0	73.9
8,000,000	24.9	18.5	1.2	84.4
9,000,000	28.1	20.9	1.3	94.9
10,000,000	31.2	23.2	1.5	105.3
15,000,000	46.8	34.9	2.2	158.1
20,000,000	62.4	46.6	3.0	211.2
25,000,000	77.9	58.8	3.7	263.7

TABLE 5.11: METAL YIELD FROM HIGH MANGANESE CONTENT NODULES AS A PERCENT OF 1985 PROJECTED DEMAND

Mining Capacity	Manganese		Nickel		Copper		Cobalt	
	Low	High	Low	High	Low	High	Low	High
1,000,000	3.6	2.7	0.3	0.3	0.01	0.01	2.1	1.9
2,000,000	7.1	5.4	0.5	0.5	0.02	0.02	4.2	3.8
3,000,000	10.6	8.2	0.8	0.8	0.04	0.03	0.3	5.7
4,000,000	14.2	10.9	1.1	1.0	0.05	0.04	8.4	7.6
5,000,000	17.8	13.6	1.3	1.3	0.06	0.05	10.5	9.6
6,000,000	21.4	16.3	1.6	1.5	0.07	0.06	12.6	11.5
7,000,000	24.9	19.0	1.9	1.8	0.09	0.07	15.1	13.7
8,000,000	28.5	21.7	2.1	2.0	0.10	0.08	16.8	15.3
9,000,000	32.0	24.3	2.4	2.3	0.11	0.09	19.0	17.2
10,000,000	35.6	27.2	2.7	2.5	0.12	0.10	21.1	19.1
15,000,000	53.4	40.7	4.0	3.8	0.18	0.15	31.6	28.6
20,000,000	71.2	54.4	5.3	5.0	0.24	0.20	42.1	38.6
25,000,000	89.0	68.0	6.6	6.3	0.31	0.25	52.7	47.8

TABLE 5.12: METAL YIELD FROM HIGH NICKEL AND COPPER CONTENT NODULES AS A PERCENT OF 1985 PROJECTED DEMAND

Mining Capacity	Manganese Low	Manganese High	Nickel Low	Nickel High	Copper Low	Copper High	Cobalt Low	Cobalt High
1,000,000	2.4	1.8	1.6	1.5	0.1	0.1	14.9	13.5
2,000,000	4.8	3.6	3.1	2.9	0.2	0.2	29.8	27.0
3,000,000	7.1	5.4	4.7	4.4	0.3	0.2	44.6	40.5
4,000,000	9.5	7.3	6.2	5.9	0.4	0.3	59.6	54.1
5,000,000	11.8	9.1	7.8	7.3	0.5	0.4	74.5	67.7
6,000,000	14.3	10.9	9.3	8.8	0.6	0.5	89.4	81.1
7,000,000	16.7	12.7	10.9	10.3	0.7	0.6	104.1	94.7
8,000,000	19.1	14.5	12.4	11.7	0.8	0.7	119.0	108.1
9,000,000	21.4	16.3	14.0	13.2	0.9	0.7	133.9	121.5
10,000,000	24.8	18.2	15.5	14.6	1.0	0.8	148.9	135.1
15,000,000	35.7	27.3	23.3	21.9	1.5	1.2	223.3	202.5
20,000,000	47.6	36.4	31.1	29.3	2.0	1.6	297.9	270.1
25,000,000	59.5	45.5	38.8	36.6	2.5	2.1	372.0	338.0

TABLE 5.13: METAL YIELD FROM HIGH COBALT CONTENT NODULES AS A PERCENT OF 1985 PROJECTED DEMAND

Mining Capacity	Manganese		Nickel		Copper		Cobalt	
	Low	High	Low	High	Low	High	Low	High
1,000,000	2.0	1.6	0.7	0.6	0.02	0.02	45.8	41.6
2,000,000	4.1	3.1	1.3	1.3	0.05	0.04	91.7	83.2
3,000,000	6.1	4.7	2.0	1.9	0.07	0.06	137.6	124.9
4,000,000	8.1	6.2	2.7	2.5	0.09	0.08	188.3	166.5
5,000,000	10.2	7.8	3.4	3.1	0.12	0.10	229.2	208.0
6,000,000	12.2	9.3	4.0	3.8	0.14	0.11	275.0	249.8
7,000,000	14.3	10.9	4.7	4.5	0.16	0.13	321.0	291.4
8,000,000	16.3	12.4	5.4	5.1	0.18	0.15	366.7	333.0
9,000,000	18.3	14.0	6.1	5.7	0.21	0.17	412.2	374.4
10,000,000	20.4	15.5	6.7	6.4	0.23	0.19	458.1	416.0
15,000,000	30.6	23.4	9.8	9.3	0.35	0.29	727.0	659.0
20,000,000	40.7	31.1	13.5	12.7	0.46	0.38	917.2	832.2
25,000,000	50.9	38.9	16.9	15.9	0.58	0.48	1145.8	1040.0

TABLE 5.14: METAL YIELD FROM DEEPSEA VENTURES' MINE SITE AS A PERCENT OF 1985 PROJECTED DEMAND

Mining Capacity	Manganese		Nickel		Copper		Cobalt	
	Low	High	Low	High	Low	High	Low	High
1,000,000	1.9	1.4	1.3	1.2	0.1	0.1	9.2	8.3
2,000,000	3.7	2.8	2.6	2.4	0.2	0.1	18.3	16.7
3,000,000	5.6	4.2	3.9	3.6	0.3	0.2	27.5	25.0
4,000,000	7.4	5.7	5.1	4.8	0.3	0.3	36.6	33.3
5,000,000	9.3	7.1	6.4	6.0	0.4	0.4	45.8	41.6
6,000,000	11.1	8.5	7.7	7.3	0.5	0.4	55.0	50.0
7,000,000	13.0	9.9	9.0	8.5	0.6	0.5	64.2	58.3
8,000,000	14.9	11.3	10.3	9.7	0.7	0.6	73.4	66.6
9,000,000	16.7	12.8	11.6	10.9	0.8	0.7	82.4	74.9
10,000,000	18.6	14.2	12.9	12.1	0.9	0.7	91.6	83.2
15,000,000	27.8	21.4	19.3	18.2	1.3	1.0	137.6	124.9
20,000,000	37.1	28.4	25.7	24.2	1.8	1.5	183.4	166.6
25,000,000	46.5	35.5	32.4	30.3	2.2	1.8	229.2	208.0

TABLE 5.15: METAL YIELD FROM HIGH MANGANESE CONTENT NODULES AS A PERCENT OF 2000 PROJECTED DEMAND

Mining Capacity	Manganese		Nickel		Copper		Cobalt	
	Low	High	Low	High	Low	High	Low	High
1,000,000	2.7	2.0	0.2	0.2	0.01	0.01	1.3	1.1
2,000,000	5.3	4.1	0.4	0.3	0.02	0.01	2.7	2.2
3,000,000	8.0	6.1	0.5	0.5	0.03	0.02	4.0	3.3
4,000,000	10.7	8.2	0.7	0.6	0.03	0.03	5.3	4.4
5,000,000	13.4	10.2	0.9	0.8	0.04	0.03	6.7	5.5
6,000,000	16.0	12.2	1.1	0.9	0.05	0.04	8.0	6.6
7,000,000	18.7	14.3	1.2	1.1	0.06	0.05	9.5	7.9
8,000,000	21.4	16.3	1.4	1.2	0.07	0.05	10.6	8.9
9,000,000	24.0	18.4	1.6	1.4	0.08	0.06	12.0	9.9
10,000,000	26.7	20.4	1.7	1.5	0.09	0.07	13.3	11.1
15,000,000	40.1	30.6	2.6	2.3	0.13	0.10	20.0	16.6
20,000,000	53.5	40.8	3.5	3.1	0.17	0.13	26.9	22.1
25,000,000	66.9	51.0	4.4	3.8	0.22	0.17	33.3	27.6

TABLE 5.16: METAL YIELD FROM HIGH NICKEL AND COPPER CONTENT NODULES AS A PERCENT OF 2000 PROJECTED DEMAND

Mining Capacity	Manganese		Nickel		Copper		Cobalt	
	Low	High	Low	High	Low	High	Low	High
1,000,000	1.8	1.4	1.0	0.9	0.1	0.05	9.4	7.8
2,000,000	3.6	2.7	2.0	1.8	0.1	0.11	18.8	15.6
3,000,000	5.4	4.1	3.1	2.7	0.2	0.16	28.2	23.4
4,000,000	7.2	5.5	4.1	3.6	0.3	0.22	37.6	31.3
5,000,000	8.9	6.8	5.1	4.5	0.4	0.27	47.0	39.2
6,000,000	10.7	8.2	6.1	5.4	0.4	0.32	56.4	46.9
7,000,000	12.5	9.6	7.1	6.3	0.5	0.38	65.8	54.8
8,000,000	14.3	10.9	8.1	7.2	0.6	0.42	75.2	62.5
9,000,000	16.1	12.3	9.2	8.1	0.6	0.48	84.6	70.4
10,000,000	17.9	14.7	10.2	9.0	0.7	0.54	94.0	78.2
15,000,000	26.8	20.4	15.3	13.4	1.1	0.81	141.0	117.1
20,000,000	35.7	27.3	19.7	18.7	1.4	1.07	187.9	156.3
25,000,000	44.7	34.1	25.4	22.4	1.8	1.34	234.9	195.4

TABLE 5.17: METAL YIELD FROM HIGH COBALT CONTENT NODULES AS A PERCENT OF PROJECTED 2000 DEMAND

Mining Capacity	Manganese		Nickel		Copper		Cobalt	
	Low	High	Low	High	Low	High	Low	High
1,000,000	1.5	1.2	0.4	0.4	0.02	0.01	28.9	24.1
2,000,000	3.1	2.3	0.9	0.8	0.03	0.03	57.8	48.1
3,000,000	4.6	3.5	1.3	1.2	0.05	0.04	86.8	72.2
4,000,000	6.1	4.7	1.8	1.6	0.07	0.05	115.8	96.4
5,000,000	7.6	5.8	2.2	1.9	0.08	0.06	144.7	118.7
6,000,000	9.2	7.0	2.7	2.3	0.10	0.08	173.4	144.3
7,000,000	10.7	8.2	3.1	2.7	0.12	0.09	202.2	168.5
8,000,000	12.2	9.4	3.5	3.1	0.13	0.10	231.4	192.5
9,000,000	13.8	10.5	4.0	3.5	0.15	0.11	260.1	216.4
10,000,000	15.3	11.7	4.4	3.9	0.16	0.13	288.4	240.5
15,000,000	23.0	17.5	6.4	5.7	0.25	0.19	458.2	382.0
20,000,000	30.6	23.4	8.8	7.8	0.33	0.25	578.2	481.4
25,000,000	38.4	29.2	11.0	9.7	0.41	0.31	723.8	602.0

TABLE 5.18: METAL YIELD FROM DEEPSEA VENTURES' MINE SITE AS A PERCENT OF 2000 PROJECTED DEMAND

Mining Capacity	Manganese		Nickel		Copper		Cobalt	
	Low	High	Low	High	Low	High	Low	High
1,000,000	1.4	1.1	0.8	0.7	0.1	0.05	5.7	4.8
2,000,000	2.8	2.1	1.7	1.5	0.1	0.10	11.6	9.6
3,000,000	4.2	3.2	2.5	2.2	0.2	0.14	17.4	14.4
4,000,000	5.6	4.3	3.4	3.0	0.3	0.19	23.2	19.2
5,000,000	7.0	5.3	4.2	3.7	0.3	0.24	29.0	24.1
6,000,000	8.4	6.4	5.1	4.5	0.4	0.29	34.8	28.9
7,000,000	9.8	7.5	5.9	5.2	0.4	0.33	40.6	33.7
8,000,000	11.1	8.5	6.7	5.9	0.5	0.38	46.4	38.6
9,000,000	12.5	9.6	7.6	6.7	0.5	0.43	52.2	43.3
10,000,000	13.9	10.7	8.4	7.4	0.6	0.48	57.9	48.1
15,000,000	20.9	16.0	12.7	11.1	0.9	0.71	87.0	72.2
20,000,000	27.8	21.3	16.9	14.8	1.3	0.95	116.1	96.4
25,000,000	34.9	26.6	21.1	18.5	1.6	1.19	145.0	120.2

ECONOMICS OF MANGANESE NODULE RECOVERY

The following information is taken from PB 226 000.

The economic value of manganese nodules has been greatly increased as a result of recent experiments pertaining to chemical processing. Prior to 1966, the cost of processing nodules to obtain manganese as MnO_2, and cobalt, nickel and copper as metals, was about $25/ton of raw nodules. With newly developed leaching techniques which remove copper, nickel and cobalt from the nodules by dissolving them away and leaving behind the manganese and iron, the cost has been reduced to $8 to $15/ton of raw nodules. As the gross recoverable value of the nodules varies between $40 and $100/ton of nodules, this processing cost is quite satisfactory.

Because the composition of nodules varies greatly from location to location on the ocean floor, a mining site can probably be chosen to yield nodules of a composition that allows disposition of all metals and products without disrupting the market as far as the total amount of these materials consumed is concerned and without over-producing any one product. Table 5.19 represents the gross values of metals in high-grade manganese nodule deposits that have been located and analyzed for their metallic contents.

TABLE 5.19: GROSS VALUE/TON OF METALS IN HIGH-GRADE MANGANESE NODULE DEPOSITS

	Minimum % by Weight	$/Ton	Average % by Weight	$/Ton	Maximum % by Weight	$/Ton
Mn	25.0	35	30.0	42	35.0	49
Ni	1.0	22	1.4	33	1.8	39
Cu	1.0	10	1.3	13	1.6	18
Co	1.3	12	0.9	36	1.5	60
Total		79		124		166

Five basic steps are involved in exploiting the mineral wealth of the seas: (1) location of deposit, (2) evaluation of deposit, (3) extraction, (4) beneficiation, and (5) transportation.

The activities and hardware used to locate deposits can be divided into three categories: ship operation, survey, and sampling. This applies specifically to seismic survey equipment, magnetometers, and gravity meters used extensively by the offshore oil industry. In addition, there is a wide variety of equipment such as sampling tools, underwater television, photographic cameras and standard oceanographic equipment applicable to seafloor studies.

After the ore has been located, it must be sampled for area and depth. Many sampling systems are available and may be classified by method of penetration. The sampled material must be removed as the ground is penetrated. This may be done intermittently with a bailer or grab, or flushed up with air or water. Once a satisfactory sample has been successfully procured standard evaluating and assaying techniques are used to determine the character, composition and value of the deposit. Since data on location and evaluation are well recorded, it might be concluded that this is not a problem, although some exploratory work and evaluation is required in the selection of the particular area to be mined before actual operation is begun.

Dredges equipped with bucket ladders, draglines, clamshells, hydraulic dredges, and air or water lifts have been successfully used in the mining of alluvial offshore deposits. Although these systems are not new for dredging, not all have been used with success in mining operations. However, all these methods have some applicability to particular cases. Figure 5.1 illustrates the three types of dredges most commonly used.

FIGURE 5.1: EXTRACTION PROCESSES

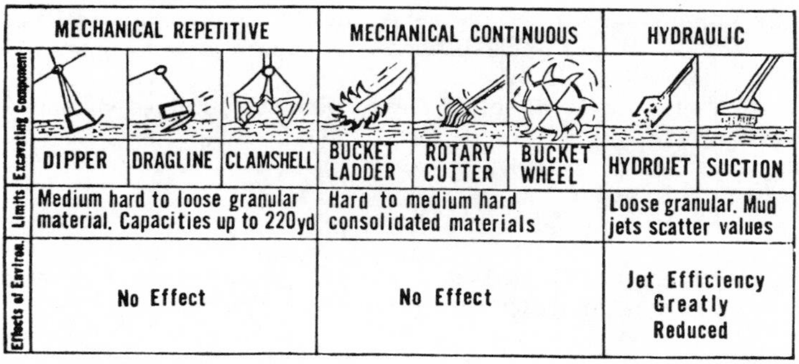

	MECHANICAL REPETITIVE			MECHANICAL CONTINUOUS			HYDRAULIC	
Limits Excavating Component	DIPPER	DRAGLINE	CLAMSHELL	BUCKET LADDER	ROTARY CUTTER	BUCKET WHEEL	HYDROJET	SUCTION
	Medium hard to loose granular material. Capacities up to 220yd			Hard to medium hard consolidated materials			Loose granular. Mud jets scatter values	
Effects of Environ.	No Effect			No Effect			Jet Efficiency Greatly Reduced	

Source: PB 226 000

Mechanical repetitive dredges are dragged along the sea bottom on a tow line. They scoop up material and are then hoisted aboard ship to determine their catch. This method is good for sampling but generally too time consuming and uneconomical for actual mining operations. Mechanical continuous dredges drill into the sea floor and flush the material up to the surface. These require a great amount of energy and have structural problems. Hydraulic dredges suck up material from the sea bottom. They too require large amounts of energy, have structural problems, and have problems with multiphase flow. Table 5.20 gives the depth applications of the presently available methods of extraction. The economically successful applications range up to 1,500 ft (460 m).

TABLE 5.20: DREDGE APPLICATIONS

Type of Dredge	Depth (ft)
Bucket Ladder	30-100
Dragline*	0-5000
Clamshell	60-200
Hydraulic	30-100
Air Lift	60-1500
Hydro Jet	0-200

*Method inefficient and economically unfeasible.

There are many processes designed for refining manganese nodules. Presently, there are a number in which the nodules are crushed, dried, leached with either hydrogen sulfide or hydrogen chloride, subjected to a separation process and then collected on electrolytic cells.

Because the areas are obviously amenable to sea-going vessels, transportation of ore presents no problem. In fact, the cost of sea transportation is far cheaper than any other, further enhancing the sea-mining operation. Just as the areas of location and evaluation have been shown not to be major problems, so also the areas of beneficiation and transportation are not major deterrents. The only remaining problem is extraction.

For any deep ocean manganese nodule mining operation the first and most significant consideration is one of economics, which is determined by

    (1)   value of the mineral content of the raw deposit as referred to in Table 5.19, and

    (2)   cost and operation of the systems necessary for recovery.

The variables necessary to determine the latter are listed in Table 5.21.

TABLE 5.21: MANGANESE NODULE DEEP-OCEAN MINING SYSTEMS; PROPERTY AND PERFORMANCE VARIABLES

    (1)   Operating Depth

    (2)   Tonnage Capacity

    (3)   Dollars/Ton of Material

    (4)   Life of System and Subsystem

    (5)   Cost of Total System and Operation

          (a)   Design
          (b)   Material
          (c)   Fabrication
          (d)   Operation

    (6)   Limitations

          (a)   Design
          (b)   Material
          (c)   Fabrication
          (d)   Operation

    (7)   Operational Reliability of Total and Individual Components

    (8)   Risk Appraisal

    (9)   Structural and Strength Considerations

(continued)

TABLE 5.21: (continued)

    (10) Effects of Current and Environment

    (11) Power Sources

POTENTIAL VALUE OF MANGANESE NODULES

The following information is taken from PB 226 004.

Although 1972 was the Centennial Year of the discovery of the manganese nodules on the deep ocean floor, it has only been in the past 15 years that any work was done concerning the possible economic use of these deposits. There are about 25 factors involved in the calculations to determine the economic value of a deposit of manganese nodules. Of these factors, the grade of the nodules is most important and, because of this, the deposits of the Pacific Ocean hold the greatest economic promise at the present time. Deposits of the nodules which are of economic interest are generally found in a broad band lying between the equator and 20°N latitude and between the North American continent and 170°W longitude. Within this area nodules containing about 5 percent combined nickel, copper and cobalt and as much as 36 percent manganese can be found. Single deposits in this general area have been found to hold a minimum of 10 billion tons of nodules.

Several techniques with which the nodules can be mined are presently under investigation. Hydraulic systems for mining the nodules have been successfully tested in 2,500 ft (760 m) of water with one such system for mining in depths of water to 18,000 ft (5,490 m) presently under construction, while mechanical, cable-bucket systems have been successfully tested in over 12,000 ft (3,650 m) of water. From standpoints of capital investment, operating cost and flexibility, the mechanical system appears to be superior.

While the nodules can be processed into salable products by a number of known industrial techniques, their unique chemical structure has made it possible to differentially leach various metals of high economic interest from the nodules. Such processes lead to relatively low plant investment and production costs. The overall production cost of winning metals such as nickel, copper, or cobalt from the nodules using a cable-bucket mining system and the differential leaching technique is now estimated to be about $0.10/lb of metal obtained. As these costs are from 15 to 25 percent of the cost of winning these metals from the bulk of the present land mines, ocean mining of the nodules appears to be economically very promising.

In addition to the lower cost of production, the nodules of the ocean floor would appear to offer many advantages in the reduction of pollution now associated with the mining and processing of metals from land mines as well as stabilizing the costs and supplies for the metals produced from them. Initially no great legal problems are envisioned in the mining of the nodules from the deep ocean floor. Once the nodules are in full-scale production, which should occur within the next 5 years, land sources of certain metals will be seriously affected.

Although 100 years have elapsed since the discovery that deposits of manganese nodules were widespread in all the oceans of the world, it has only been within the past 15 years that they were recognized as a major potential economic source of industrial metals for the world population. This realization came about through a study conducted in 1957-1958 by the Institute of Marine Resources of the University of California after a large haul of nodules, rich in cobalt, was dredged from Tuamotu Escarpment, just east of Tahiti. The haul was raised by scientists from the Scripps Institution of Oceanography as part of the 1957 International Geophysical Year program. The results of that study were quite favorable as to the technical and economic aspects of mining and processing the nodules. The published results of that project have sparked the investment of almost $100 million by natural resource companies of the world in the development of the nodules as a source of metals.

It is particularly fitting, in these times of hypertensive environmental awareness, that the manganese nodules of the deep-sea floor should be under such intensive investigation because (1) there will be no measurable environmental damage done in the mining and processing of these nodules; (2) the full-scale development of these deposits as a source of industrial metals will allow society to close many of the sulfide mines on land which are presently a substantial source of air and land pollution; and (3) due to the unique physical and chemical structure of the nodules, with their large and chemically reactive specific surface areas, there is some indication that the nodules may be quite useful in greatly reducing pollution of the atmosphere from other operations such as power production and automobile exhaust emissions.

What other applications may be found for this gift of the sea remains to be seen when the nodules are brought under intensive investigation of all of their facets.

There are about 25 factors involved in the calculations used to determine the economic value of a deposit of manganese nodules; the more important ones include grade of nodules, concentration of the nodules per unit area of ocean floor, size distribution of the nodules, physical characteristics of the associated sediments, depth of water, distance to port or process facility, topography of the ocean floor in the deposit area, and weather in the deposit area.

Of these considerations, the most important factor is the grade of the nodules. Because of this consideration, it is the nodule deposits of the Pacific Ocean which are of the greatest interest at this time. More specifically, it is the nodules in a band between the equator and 20°N latitude and between the North American continent and about the 180° line of longitude which are of greatest interest, for it is in this area that the nodules of highest economic value are found.

Exploration of the nodule deposits is generally accomplished, for economic purposes, by scanning the deposits with a television camera or still camera and periodically sampling with some dredging device, usually a bucket which is also used for tonnage sampling of the deposits for process development work. Some companies use a series of free-fall photographic and grab sampling devices with sample points spaced about 1 mile apart through a deposit area. The cost of a sophisticated exploration program to find and block out say 10 million tons of the nodules would be about $300,000. The costs of such a program, however, can vary greatly depending on the specifications for sample-point spacing.

Off the south coast of Mexico 100 miles there is quite a high concentration of large nodules, approximately 300,000 tons per square mile at a depth of about 12,000 feet. About 1,000 miles southeast of Hawaii in 15,000 feet of water there are perhaps 15,000 tons of nodules per square mile. About 500 miles northeast of New Zealand, there is an area of what one might call 100 percent concentration, the nodules are practically touching one another. We can sample the deposits or dredge approximately 500-ton lots for process development work; the nodules have been dredged with this type of technique for the past 10 years. A larger system is the 8-ton bucket which was used to depths of 20,000 feet. Some of the techniques developed 15 years ago were so good that they are still being used.

A conservative estimate indicates that there are several hundred billion tons of mineable nodules in the high-grade areas of the Pacific Ocean. The highest grade of nodules yet discovered and reported is found in a deposit about 1,000 miles north of Samoa. Nodules from this deposit will assay about 1.9 percent of nickel, 2.3 percent of copper, 0.2 percent of cobalt and 36 percent of manganese on a dry weight basis. Deposits of the nodules can be found in other areas of the ocean which assay as high as 2.6 percent of cobalt or 55 percent of manganese.

In general, the chemical composition of the nodules is very uniform over large lateral distances of the Pacific; however, the concentration of the nodules can vary markedly throughout any given deposit. The highest concentration of nodules, excluding the crustal deposits, presently known is about 100 kg/m^2 of sea floor which would work out to abut 300,000 tons/square mile. An average concentration in a deposit considered for mining would probably be in the range of 30,000 to 75,000 tons/square mile of sea floor. In general, the average size of the nodules is about 4 cm; however, within a given deposit, this size range may vary from 1 to 20 cm.

Nodule mining companies seem to be interested only in the monolayer of nodules at the surface of the sea-floor sediments. Although nodule beds can be found at a number of horizons down the sediment column, it is not thought that it would be economic to mine and process the gangue sediments to secure the buried nodules.

The fact that the nodule deposits exist only as a thin surficial monolayer measurably complicates the design of an effective mining system, as very large areas of the ocean floor must be swept over to allow recovery of the nodules at economic production rates. Although it is possible that the nodules may be piled up in depth in some areas of the ocean floor, generally the devices used to sample the sea-floor nodule deposits would not disclose this fact. If such deposits had been found initially, the mining of the nodules would have been an accomplished fact many years ago. Also, it is not thought possible to mine the crustal manganese deposits of the sea floor due to the difficulties of breaking these crusts free from their solid attachment to sea-floor bedrock.

Although deposits of manganese nodules can be found in almost all depths of water in the ocean (they can be found in 6 feet of water in some Scottish Lochs), only those lying below about 3,000 m of water are presently being considered as economic to mine. The higher grades of nodules are generally found in depths of water ranging from 4,000 to 6,000 m.

On a Pacific Ocean-wide basis, it has been estimated that there are some 1.5 trillion tons of the nodules presently at the surface of the sea floor and that the nodules are forming in this ocean at the rate of about 10 million tons/year. On the basis of an average composition, ocean-wide, Table 5.22 lists some statistics on the amount of various metals contained in the nodules of the Pacific Ocean and the rate of accumulation of the metals in these deposits.

TABLE 5.22: RESERVES OF METALS IN MANGANESE NODULES OF THE PACIFIC OCEAN

Element	Amount of Element in Nodules (Billions of Tons)[a]	Reserves in Nodules at Consumption Rate of 1960 (Years)[b]	Approximate World Land Reserves of Elements (Years)[c]	Ratio of (Reserves in Nodules)/(Reserves on Land)	U.S. Rate of Consumption of Element in 1960 (Millions of Tons/Yr)[d]	Rate of Accumulation of Element in Nodules (Millions of Tons/Yr)	Ratio of (Rate of Accumulation)/(Rate of U.S. Consumption)	Ratio of (World Consumption)/(U.S. Consumption)
Mg	25.	600,000	L[f]	–	0.04	0.18	4.5	2.5
Al	43.	20,000	100	200	2.0	0.30	0.15	2.
Ti	9.9	2,000,000	L	–	0.30	0.069	0.23	4.
V	0.8	400,000	L	–	0.002	0.0056	2.8	4.
Mn	358.	400,000	100	4,000	0.8	2.5	3.0	8.
Fe	207.	2,000	500[e]	4	100.	1.4	0.01	2.5
Co	5.2	200,000	40	5,000	0.008	0.036	4.5	2.
Ni	14.7	150,000	100	1,500	0.11	0.102	1.0	3.
Cu	7.9	6,000	40	150	1.2	0.055	0.05	4.
Zn	0.7	1,000	100	10	0.9	0.0048	0.005	3.5
Ga	0.015	150,000	–	–	0.0001	0.0001	1.0	–
Zr	0.93	+100,000	+100	1,000	0.0013	0.0065	5.0	–
Mo	0.77	30,000	500	60	0.025	0.0054	0.2	2.
Ag	0.001	100	100	1	0.006	0.00003	0.005	–
Pb	1.3	1,000	40	50	1.0	0.009	0.009	2.5

[a]All tonnages in metric units.
[b]Amount available in the nodules divided by the consumption rate.
[c]Calculated as the element in metric tons. (From U.S. Bureau of Mines Bulletin 556).
[d]Calculated as the element in metric tons.
[e]Including deposits of iron that are at present considered marginal.
[f]Present reserves so large as to be essentially unlimited at present rates of consumption.

An interesting calculation indicates that many of the industrially useful metals are accumulating in the nodules at rates which exceed the present world consumption of these metals. Once the nodules are being mined, the mining industry would be faced with the problem of working a deposit which forms at rates greater than it can be mined.

Generally, on land, mineral deposits are considered as depleting resources; but in the ocean it is found that many of the mineral deposits of economic value, including deposits other than manganese nodules, are actually forming at the present time at rates which greatly exceed present world consumption of those minerals.

This observation is of academic value only, as the present reserves of metals in the existing deposits would supply the world population for thousands of years even at per capita consumption levels of those in the highly developed nations of the world. Of course, one of the major advantages in considering the nodule

deposits of the ocean as a source of industrial metals is that they are open to
any nation or population of the world that might wish to recover them. Continental mineral deposits, in a number of cases, are closed off to all but a hand-
ful of potential customers.

The sea-floor nodules can be reduced to salable nickel, cobalt and copper metals,
or manganese ore by a number of standard industrial techniques. Generally,
these techniques involve the dissolution of all valuable metals in the nodules
and the differential precipitation or separation of the metals from solution.
These processes generally lead to high plant capital costs, in the order of $50
to $100/annual ton of nodule capacity, and high operating costs, in the order
of $20 to $30/ton of nodules processed. Rather high recovery efficiencies, however, can be achieved with these processes in the order of 90 to 95 percent of
the contained metals in the nodules.

Because of the unique chemical structure of the nodules (with the bulk of the
copper, nickel and cobalt metals being contained therein as ions loosely attached
to the surfaces of the manganese and iron hydroxide minerals), it has been found
possible, not only to leach differentially these metals from the nodules, but to
do it in a way which prevents any appreciable amount of the manganese or iron
from going into solution.

Such a process, which has been discovered and developed at the University of
California at Berkeley, and which can be essentially similar to oxide heap leaching processes, leads to very low process plant capital costs and operating costs.
With such a process the plant capital costs can be expected to be in the range
of about $10 to $20/annual ton of nodule capacity and the operating costs
would be in the range of $5 to $10/ton of nodules handled. While recovery
efficiencies with this differential leaching process are comparatively low, in the
order of 60 to 80 percent of the contained metals, it does minimize capital investment and maximize return on investment.

The use of a hydraulic system to mine the nodules, because it is not possible to
operate this system at low capacities, in general indicates a total overall capital
investment in the mining system of at least $30 million. Coupled wtih a differ-
ential precipitation process for processing, which would indicate a process plant
investment cost of about $75 to $100 million to handle one million tons of the
nodules per year, the total capital investment of such a nodule mining and processing system can be expected to be at least $135 million.

Such a venture could produce about $67 million worth of products annually
operating at an overall recovery efficiency of 90 percent and assuming the manganese-iron by-product left over after the other metals are removed is acceptable
as a metallurgical-grade manganese ore. The cost of operating this system is
estimated at about $39.3 million/year, yielding a gross operating profit of about
$27.7 million/year and allowing a gross rate of return on investment of about
21 percent/year. Such a system would be essentially competitive with land mining operations to produce these metals.

The use of the Continuous Line Bucket system of mining, coupled with a differential leaching process, to produce the nodules at a rate of 1 million tons
per year would appear to involve a total capital investment in mining and processing facilities of about $20 million. This operation would produce about

$52.1 million/year worth of products operating at an overall recovery efficiency of 70 percent. The operating costs of this system are estimated at about $18.3 million/year yielding a gross operating profit of $33.8 million/year for a return on investment rate of 169 percent/year. Some of the statistics of such operations are shown in Table 5.23.

TABLE 5.23: OVERALL ECONOMICS OF TWO SEA-FLOOR NODULE MINING SYSTEMS[a]

ITEM		Hydraulic System		CLB System	
		Capital Costs ($)	Oper. Costs ($/Y)	Capital Costs ($)	Oper. Costs ($/Y)
1) Preliminary Investigations:		1.0[b]	—	0.1	—
2) Nodule Deposit Exploration:		0.3	0.3	0.3	0.3
3) Mining System:		30.0	10.0	2.0	2.0
4) Process System (Diff. Ppt.)		75.0	20.0	—	—
Process System (Diff. Leaching).		—	—	10.0	10.0
5) Transport of Nodules (Assuming Chartered Vessels and 4,000 Mile Roundtrip, Deposit-to-Process Site:		—	4.0	—	4.0
6) Operating Capital and Misc. Costs:		28.7	5.0	7.6	2.0
Total Estimated Capital + Operating Costs:		135.0	39.3	20.0	18.3
Value of Products Recovered[c]:	($/Y)	67.0		52.1	
Gross Net Operating Profit:	($/Y)	27.7		33.8	
Gross Rate of Return on Inv.:	(%/Y)	21		169	
Cost Per Pound of Co, Ni, and Cu Metals Produced:	($/lb)	0.23		0.12	
Cost Per Ton of Mn-Fe Ore Produced[d]:	($/T)	84.50		45.80	

[a]On a basis of 1 million dry-weight tons of nodules per year assaying 40% Mn+Fe, 0.2% Co, 1.6% Ni and 1.4% Cu.
[b]$ and $/Y figures quoted in millions of dollars.
[c]Assuming process recovery efficiencies of 90% for differential precipitation process and 70% for the differential leaching process. Mn-Fe in nodules assumed to be produced in the form of a manganese ore valued at about $35/ton. Value of other products assumed at $2/lb for cobalt, $1.20/lb for nickel, and $0.50/lb for copper.
[d]Production costs distributed: Co+Ni+Cu — 30%; Mn+Fe ore — 70%.

As indicated in Table 5.23, the cost of handling the manganese-iron portion of the nodules would appear to be somewhat greater than the value of the product produced indicating that there would be a substantial gain in the profitability of the venture if the copper-nickel-cobalt metals could be taken from the nodules on shipboard at the mining site and the remainder of the nodules returned to the sea floor to seed the mined out areas for formation of nodules for future generations.

The economics of any venture to mine and process the deep sea nodules, of course, can be measurably improved by increasing the scale of the operation and by producing some of the other metals found in the nodules such as molybdenum, lead, zinc, zirconium, cerium, etc. Also, it would appear, that initially the mining of metals from the deep ocean will be a tax-free situation because the mining and processing facility can easily be set up in one of a number of tax-free nations. Companies from high-tax nations would simply be factored out of ocean mining for the nodules as they would not be competitive if forced to continue paying their high tax rates.

As a source of revenues for development in the poorer nations of the world, the manganese nodules will never meet expectations. The total revenues generated from such operations, even if present land-derived metal prices were maintained for ocean products, would not provide more than a few cents per capita for the people of the less developed nations.

Multiplying the total tonnage of the nodules presently speculated to be at the surface of the sea floor by the $50 to $100/ton of metals which can be won from the nodules, a grand figure of several hundred trillions of dollars is arrived at. This is then assumed to be available for the gathering. This, however, is an illusion. The developing nations will stand to gain infinitely more by the availability to them of markedly less costly basic materials produced by themselves.

The total potential economic value of the nodules in the ocean can be calculated, but it would be a meaningless value. When full-scale production of metals from the nodules is achieved, markets and prices for the metals will change markedly, with the prices to the consumer falling greatly and the markets expanding greatly. With low-cost nickel, industry would be able to use stainless steels in place of carbon steels for automobiles and structures. The result will be a great gain in overall metal utilization efficiency due to the corrosion resistance of this material.

Stated simply, the manganese nodules of the deep sea represent an apparently less expensive and essentially inexhaustible source of many important industrial metals for all populations of the world. They also represent a new and exciting investment opportunity which will not last more than a few years after initial production of metals from the nodules is achieved.

They will, of course, also represent a means of measurably reducing pollution of the atmosphere by permitting the closing down of many pollution-prone sulfide mines on land. Moreover, it may be possible to use the nodules for the removal of sulfur dioxide from the gases of power plant stacks, thereby permitting power companies to burn cheaper, high-sulfur fuels. Because of the very high surface areas of the nodules (in the range of 100 to 300 m^2/g of nodule material) and because of the highly reactive surfaces, the nodules may also prove important as general catalytic agents. One application could be to convert unburned hydrocarbons in automobile exhausts, a prime cause of smog, to harmless carbon dioxide.

LAND-BASE REQUIREMENTS OF NODULE MINING

The following information is taken from PB 218 948.

Land-base needs for deep-ocean mining are very closely related to site selection; i.e., determining the most economical location of the nodule processing plant support activities. Key considerations of site selection are the availability of energy and the need to provide for effective environmental protection. This will include the plant needed to extract the metals contained in manganese nodules since it will probably be a large chemical processing facility. In order to provide a proper frame of reference, the principal elements of the entire ocean mining operation listed below will be briefly reviewed.

Mine site and ore body
Mining machine(s) and vessel(s)
Ocean transport - ore; product
Land transport - ore; product
Process plant - sea; land
Logistics support
Administrative support - marketing
Research and exploration

The principal elements of an ocean mining operation include first a mine site and an ore body. Because nodules are present, for example, in the Pacific Ocean, it does not mean that they represent an ore body. From the miners' viewpoint, a great deal of effort and information is required before a deposit or occurrence can be considered an ore body. At this stage in the development of the resource, according to data published by the scientific community, there are not as many mine sites or ore bodies available as has been assumed.

The next element is the mining system. The mining machine gathers and collects the nodules and brings them to the sea surface from the sea floor. The ocean transport elements carry the ore to shore from the mine site. Also because of the large quantities of products which are extracted, and which will undoubtedly be marketed worldwide, ocean transportation of the product will be required.

Land transport is another major element. If the plant is located in the continental United States, large quantities of ore may be transported by land from a marine terminal to the plant site. In addition, some of the products will have to be distributed to markets by land transport modes.

The process plant is needed to beneficiate and refine the ore. There are many approaches to nodule ore processing including location of the plant at sea on a ship or floating platform. The natural location for the initial plant at least is on land, and it is probable that the first nodule plants will be located on shore.

Logistics support is another major item. A large marine-chemical plant operation will require a major back-up support complex including the logistics support for the ocean mining operation and the process plant. This must be complimented by administrative support functions and product marketing. Finally research and exploration must constantly be in progress, research in process development and exploration to find better mining sites. The principal land elements include a marine terminal, the mining operation supply system, the process plant, and associated administrative support.

The marine terminal will be examined with regard to deep water, land, ore unloading and stacking, mining ship fuel/water, product transportation, and reagent delivery. For a large-scale operation which contemplates mining at least a million tons of nodules a year (assuming nodules are to be brought to shore for processing), the size ships that are required to do this economically, will range between 35,000 to 60,000 tons deadweight. These ships require draughts of from 35 to 40 feet full load, and perhaps will be between 600 and 700 feet in length. Therefore, a deep-water port will be required. This is not the same type of deep-water port that is needed by the super tankers (LCC) to transport petroleum. Ports with water depths of up to 40 feet are available in many areas throughout the United States. There are not, however, enough of them on the mainland or in Hawaii to provide maximum flexibility needed in site selection.

The decision of where to locate the marine terminal is of major importance and each candidate must provide the complete facilities needed to handle the large ships contemplated. To permit an understanding of what the traffic might be for such a deep-water port for one processing plant, and assuming the terminal is in the State of Hawaii, the following example is provided. The round trip to the mine site might be 3,000 miles. A one-million-ton-a-year operation will require 20 round trips a year per 50,000-ton ship. Assuming 12 days per round trip, this means that one ship can handle the operation and it would be coming to port about once every two weeks to unload the unprocessed nodules for a one-million-ton-per-year operation.

For two-, three-, and four-million-ton-per-year operations, the traffic would increase proportionately. The marine terminal requires a fairly large waterfront and pier facility and also land facilities for stacking and storing the nodules as they are unloaded from the ore carrier. Included at the facility will be some type of large, high-capacity unloading equipment for handling the ore. The ocean mining company may elect to build its own private marine terminal, depending upon the quantity of material handled. If there is enough nodule ore to keep the terminal occupied on a dedicated basis, then the ocean mining company might include the construction and operation of the terminal in its system.

On the other hand, if it is just a one-ship operation, a minicipal or public bulk unloading terminal may be sufficient. The process plant is best located adjacent to the marine terminal to reduce land transportation distances and costs. Indications are that for the size plant which will be needed, the land requirement will range between 100 and 300 acres. The required unloading capacities at the terminal will be in the 2,000 to 3,000 ton-per-hour range.

The mining ship will be operating at sea, undoubtedly all year around, and will have to be refueled at sea. The refueling can be accomplished from the ore transport ship by carrying fuel out to the mining ship. As the transfer of nodules or ore takes place, the mining ship is simultaneously replenished with fuel. Refueling needs for the mining ship may require the installation of appropriate oil-storage capacity at the marine terminal. This might be on the order of perhaps 40,000 to 100,000 tons per year of fuel oil and will require a small fuel oil tank farm.

Also if some type of washing or separation of nodules is accomplished on the mining ship, fresh water may have to be taken out to the mining ship. The evaporating capacity of the mining ship may not be large enough to provide the needed fresh water. A study of product transportation needs indicates that a large percentage of the metals produced, particularly if produced in the Hawaiian Islands will be transported to market by sea. For a plant processing one million tons of dry nodules per year, for example, approximately 250,000 tons of manganese, 12,000 tons of nickel, 10,000 tons of copper, and 2,300 tons of cobalt can be produced. The marine terminal, if adjacent to the plant, can provide convenient dockage for the product transport ships.

Associated with the plant and with the marine terminal is the delivery of required reagents. These will be needed to operate the chemical process plant. There are many basic processing approaches for the extraction of the metals contained in the ore: hydrometallurgical, pyrometallurgical, etc. Each in turn will require several types of special reagents. The availability of reagents locally is of major importance in site selection. In the State of Hawaii, unfortunately, most of the needed reagents are not readily available. The next segment of the land operation is the supply base involving pier and docks, office (port captain and staff), warehouse (spares, supplies, etc.), supply ship/crew boat and survey ship.

Once the mining operation is underway at sea, a base for support of the marine operation will undoubtedly be required. This base need not be located at the marine terminal or at the plant site. The principal elements include piers and docks for a supply ship. The supply ship will transport personnel to and from the mining ship for changeover of operation personnel, and also will transport the required stores and spares.

In addition, space ashore will be needed for the port captain and his staff and for large warehouse facilities for repair parts and supplies for the mining ship. A fairly high-speed crew boat will probably be needed to transport the personnel. Possibly accompanying the mining ship will be a survey ship, needed to delineate the mine site incrementally as the mining will take place over many years. The essential elements of the process plant are: land (stockpile, plant, storage), plant construction (materials, cost), energy (electricity, gas, oil), water, reagents, labor, and waste disposal.

A large tract of land will be required. The land requirements depend upon the size of the plant; i.e., the quantity of nodules to be processed and land needs can be between 100 to 300 acres or more. Major factors that must be evaluated in site location analysis include the local cost of plant construction and materials. This information will be needed by an ocean mining company in conducting its site selection study. The availability of this information from State representatives would be very helpful in evaluating siting a plant.

Electrical energy capacity for a nodule processing plant can be on the order of from 50 to 200 megawatts, depending upon size and type of plant; i.e., the tons of nodules to be processed and the type of process selected. The possible availability of geothermal energy, for example in Hawaii, sounds very exciting and may be a way whereby Hawaii could provide clean energy at prices comparable to fossil-fuel produced electrical energy now available on the mainland.

Other energy or reagent requirements may include gas and oil or coal. The process plant will also require water. Large quantities of water will be needed for process and cooling uses. Process water probably can be recirculated. Depending upon the process used, cooling water requirements can be anywhere from 300,000 to 600,000 tons per day. Depending on the process used, reagents can include acids, caustics, chemicals and salt (brine). The brine requirement may be satisfied in conjunction with a geothermal plant. Brine, a potential waste of a geothermal plant, could be useful as a reagent for the process plant. Labor requirements will include construction labor and operational labor. For a plant located in Hawaii, it is expected that the major skills are available to build a complex chemical plant. However, labor costs are known to be high. This type of information will also have to be provided for evaluation of a potential site.

Another very important item is waste disposal. From an environmental point of view the use of a chemical process for extracting the metal values in the nodules will be very beneficial because it can be designed to recycle reagents. Anticipated discharges into the atmosphere, water vapor and carbon dioxide, are not considered to be pollutants. If liquid waste is produced, it will be sufficiently clean to permit harmless discharge into the ocean. However, the question of solid-waste disposal is going to be critical. For instance, 100 pounds of dry nodules which are to be processed only for copper and nickel, will contain approximately one pound of gas and water vapor, about 3 pounds of product and about 96 pounds of solid waste.

This means that in a million tons of nodules there can be 960,000 tons of solid waste in need of disposal. To provide a physical feel for this, the amount of material might cover 10 acres 3 to 4 feet high. The solid wastes are oxides and silt, and are very fine, probably 100 to 300 mesh, and may cause a major waste problem. Solid-waste disposal will have to be given very careful thought in planning for the nodule plant.

The natural thing of course would be to return the waste to the sea; it came from the sea, and perhaps it can be returned to the sea via the transport ship returning to the mining site. Another approach is to not bring the wastes to the land at all; try, if possible, to leave them in the sea at the mine site by processing at sea. This approach also has many major problems.

On the other hand, a plant that processes the nodules for the manganese as well, is in a slightly better position because once the oxides are removed (in particular the manganese oxide) the resulting residue is primarily clean silt and has water-holding capacity. It probably can be spread out over the lava fields of Hawaii and in a very short time could be used to cultivate agriculture products. It also could be used in clean land fills. These certainly are viable solutions.

A final item of importance is general and administrative support, that is: taxes (local, state), insurance, labor relations, wage rates, community considerations (cost of living, schools), climate, communications, and government (local, state). The state and local tax situation must be evaluated. In addition, what sort of helping hand, if any, will the local government give to businesses looking to the siting of plants or bases in the area under its jurisdiction? Other factors include: insurance; labor relations history; wages (relative wages compared with other areas); cost of living; schools; food; climate; local communications and world-wide communications; and the general attitude of local and state government towards the establishment of an industrial complex in the area.

VARIABILITY OF PACIFIC OCEAN NODULE DEPOSITS

The following information is taken from PB 226 012.

The distribution and population of nodule deposits on the deep sea bed have been extrapolated from widely spaced still photogrpahs of the sea floor. Continuous TV scanning of the ocean bottom has shown that the distribution is exceedingly variable, even over relatively short distances. TV video tapes of actual nodule deposits of the Pacific Ocean at depths of 3,000 to 6,000 m demonstrate this variability.

The mineral exploration program of Deepsea Ventures, Inc. consists of four major effots: (1) to locate nodule deposits of potential commercial interest; (2) to conduct preliminary surveys of these deposits and delineate their extent; (3) to conduct more extensive surveys of these deposits and evaluate each as possible ore bodies; and (4) to develop a detail mining plan for acquisition of the ore.

During phase 1 and 2 efforts, the principal objective is to locate ocean sites at which there is a high possibility of an economic deposit. Phase 3 and 4 surveys are conducted in much greater detail at carefully selected sites determined from results of the previous efforts. The decision as to where to conduct preliminary surveys is a difficult one, as the cost of operating an oceanographic ship is relatively high. Planning of the survey cruise requires careful analysis and evaluation of available data on nodule occurrence. Much of this fundamental data is acquired by the scientific community. Utilizing this and other information, a research team is dispatched to study the areas of greatest potential.

Evaluation criteria are used to determine the merits of a deposit. These include: bathymetry; bottom topograpy; nodule population and concentration; assay; local sediment soil mechanics; and the local marine environment. The most valuable tool is the deep ocean TV system which permits continuous scanning of the sea floor. Data acquired on each phase 2 cruise is carefully analyzed and utilized in the economic assessment model of a mining system to evaluate its potential as a commercial mine.

JAPANESE PROGRAMS FOR MANGANESE NODULE EXPLOITATION

The following information is taken from PB 218 948.

Japan is an island country, and its limited land resources cannot house its rapidly growing industry. Table 5.24 compares world copper and nickel consumption in 1960, 1965, and 1969. The need for copper and nickel has increased in Japan; nickel consumption has grown about 3.9 times in the past 9 years.

TABLE 5.24: WORLD CONSUMPTION OF COPPER AND NICKEL

Metal	1960	1965	1969	Growth Rate, percent
	- - - - - - - - - - -(10,000 tons)- - - - - - - - - - -			(1960-1969)
Copper	320	433	807	10.8
Nickel	20	28	78	16.3

Most of Japan's copper and copper ore is imported, and all of her nickel and nickel ore is imported. For example, about 2.7 million tons of nickel ore was imported from New Caledonia Island in 1969. While nickel content in ore has decreased, the price has increased.

In examining future nickel, copper, cobalt, and manganese resources, deep-sea nodules have great potential interest to Japan. Manganese nodules can absorb a great amount of sulfur dioxide, nitrous oxide, etc. and they can be very effective in controlling gas pollution. Japanese industry produces severe pollution. Power plants, especially, use oil which is rich in sulfur. Antipollution machinery is necessary; however, it is difficult to find an inexpensive and effective method. Thus deep sea nodules will be of interest to Japan not only for metal resources, but also as antipollution material.

The deep-sea floor is flat and wide, and nodule resources are on the sea floor rather than under it. The 5,000-meter depth is not as far as the moon: it is just 4 hours for a man walking. Using the CLB conveyor between the sea surface and the sea floor, it proved easy to carry nodules to the sea surface after some improvements were made. If from 500 to 1,000 tons of nodules per day from 5,000 meters were dredged using a 20,000-ton Japanese ship, daily costs would be:

Ship charter	$5,000
CLB ($1 million; 3 year depreciation)	1,000
Other fees	2,000
Total	$8,000 per day

If it takes two days to collect 1,000 tons including transportation and contingency days, it will be $16 per ton; thus the current cost estimate is $15 to $30 per ton. Of course, it will be much less expensive when deposits are near port facilities and are shallower, such as those near the island of Kauai, and the cost of the CLB will decrease with increased production.

REQUIREMENTS IN MANGANESE NODULE EXPLORATION

The following information is taken from COM 73 11947.

In the exploration for deep sea manganese nodules, many environmental factors must be considered and technology must be developed to overcome the imposed restraints. Exploration in the practical context must be defined as the process of determination of the benefit cost ratio of any particular mining operation. It includes, therefore, not only the prospecting effort, which substantiates the presence of mineral values, but, also, the deposit characterization and environmental monitoring which will permit quantification of the benefits and the costs in real terms.

The benefits are obviously the utilization of the resource and all the externalities associated with having the resource. The cost is what it takes to provide the benefits in terms of money, technology and effects. Thus, exploration is a measuring process carried out for the purpose of making sound decisions on future exploitative processes. Two questions, therefore, must be asked: (1) What must be measured? and (2) How are these measurements made? Many data are required to characterize a sea-floor mineral deposit. Ferromanganese deposits are widely distributed throughout the world ocean, and it is evident that they exhibit wide variations in tenor and in environmental setting, even over moderately short distances.

Thus, the characterization of deposits is dependent on a thorough knowledge of the geological, geotechnical, oceanographic, meteorological and ecological parameters, and the measurement of each of these is an essential part of any nodule exploration program. Expansion of the data is, therefore, of paramount importance, and it must be done prior to commencing even pilot mining operations.

Geological

A more basic data requirement is that of the deposit and its sea floor environment. It is necessary to understand the topographic features of the deposit in both large and small scale in order to evaluate methods and develop costs for exploitation. The value of the ore is determined from its composition which may vary in relation to the geochemistry of the underlying sea floor and the superjacent waters.

Costs of processing will also be dependent on the texture and mineralogy of the ore and gangue materials, and leads for further prospecting will no doubt be engendered by an understanding of the ore-forming process. Sampling is an integral requirement of the geological data base though many improvements in remote sensing are obviating some of this need.

Through the use of bottom photography, rock sampling (if exposures are present) and both grab samplers and core samplers, the geological-sedimentological regime at the potential site should be sampled and described. In regard to the sedimentary matrix which is, in fact, the principal gangue of the exploitable nodules, the mining operator needs to know both the physical as well as the chemical characteristics of the sediment. To acquire such data requires that either actual samples be taken or that some in situ sensing system be employed which, once calibrated, will provide the much-needed physical and chemical information. To solve this problem alone requires research and development efforts in deep-sea sensing and instrumentation. Although empirical relationships are known to exist between certain engineering or geotechnical properties of bottom sediments and such sedimentological parameters as grain size, specific surface area, sorting and even gross mineral content, there is a need to collect physical samples of both the sediment and the nodules.

Geotechnical

The marine mining environment consists basically of the air-sea interface, the water zone, the sea-floor and the subbottom. Characterization of this environment and an understanding of its interaction with the various mining systems is one of the key requirements in developing successful marine mining technology. Depending on the type of deposit (dissolved, unconsolidated or consolidated) and its geographic location, each deposit will have a different environmental setting that must be described in order to define the specific interactions between the environment and the mining system. In evaluating the economic feasibility of a commercial offshore placer operation, there are two main factors to be considered: (1) the value of mineralization available for extraction and (2) the cost of delineation and mining. The latter factor represents a major unknown which can only be answered by extensive analysis of the mining system components and their interaction with the environment. Successful marine mining systems will be possible only by achieving better operational control and a better understanding of the engineering properties of the deposit material being mined. Thus, characterization is basic to successful mining systems design.

The most important parameters affecting dredge performance and the most difficult to measure are those describing the material to be dredged. The usual method of determining dredging progress is to conduct hydrographic surveys before, during and after dredging and compute the volume of material removed. For control purposes during the actual dredging operation, continuous monitoring is required in order to evaluate the dredging equipment and the operating procedures and to ensure optimum productivity. This information can best be obtained by the use of performance monitoring instruments.

The importance of subbottom characteristics is equally important in evaluating other subsystem components. Great improvements could be made if the effect of soil parameters on penetration rate, core recovery and sample reliability were better understood. Techniques are being utilized to determine true in-place density in conjunction with routine deposit delineation. This can be accomplished by controlled coring using very thin-walled tubes for short penetrations, by in situ penetration measurements based on a preestablished penetration-density index or by indirect means such as acoustic or nuclear-density measurements. The design and performance of other subsystems are also affected by the deposit characteristics.

Transportation, beneficiation and waste disposal techniques require adequate information concerning the grain size and sedimentation characteristics of the excavated material as well as its consistency and plasticity grading in the disturbed state.

Problems with bottom slope stability will also need solutions in order to prevent contamination of the mining areas or burial of bottom-supported equipment as a result of slope failure. The use of equipment on the bottom will require knowledge of the sea-floor-bearing capacity. Solution of these problems requires knowledge of sediment shear strength and compressibility. Sonic velocity characteristics of the deposit material are needed for the interpretation of subbottom profiles.

Oceanographic

Measurement of physical and chemical oceanographic variables is a necessary adjunct of deep-sea mineral exploration for a number of reasons. The mineral content of the deposits may be influenced by the nature of the water interface; operations will be strongly influenced by the dynamics of the water mass, and the biomass is water dependent in a variety of ecological niches, any of which could be affected by disturbances caused by mineral exploitation. The state-of-the-art for many of these measurements, particularly on a widely dispersed, synoptic basis, is not sufficiently advanced for mining use. There are, for example, few reliable measurements of bottom currents in the deep ocean. Most estimations have been made from the configuration of sediment ripples or the amount of sediment coating on jutting rocks or nodules as observed in photographs or from theoretical considerations. Cyclical or seasonal measurements are virtually nonexistent.

The art of measuring sea state is a little more advanced, but prediction is not yet fully reliable. The geochemistry of ocean waters and the identification and measurement of particulates is again in a primitive technological state with regard to the deep sea.

Programs such as the Geochemical Ocean Sections Study (GEOSECS) of the IDOE will do much to improve capabilities in this regard, but the very vastness of the deep seas will make significant coverage a formidable task.

A review of the available information suggests that the measurement of oceanographic variables should be by automatic, self-recording in-situ sensing systems.

Meteorological

Most data on weather and climate are required in order to evaluate available working time in areas adversely affected by weather conditions and also to reduce the risk of loss of life and property. The weather is mostly a surface phenomenon, and, because of its very substantial impingement on operations at sea, good meteorological data are of paramount importance.

All this could be obviated in the deep sea, however, by placing the operations in a submerged mode and making use of the predictable climate of the sea-floor environment. Thus, prior to initiating a major meteorological program, some serious study should be given to the possibility of placing parts of any nodule mining system on the deep sea floor or, at least, to submerging parts of the system beneath the surface of the sea.

Ecological

The collection of data on the life of the deep sea and the understanding of environmental change is a complex and formidable task. The basic requirement for evaluating the influence of a mining is a knowledge of existing life in the water and on the sea floor. The food chain cycle has to be determined from the surface layers of primary photosynthetic production to the detritus dependent creatures in and on the bottom.

Natural changes must be recorded and the effects of artificial change predicted and superimposed. The ecological data base is virtually unknown in the deep sea and new techniques will have to be developed in order to obtain the necessary data.

In the exploitation of ferromanganese nodules, three major types of disturbance can be anticipated, according to the present state-of-the-art. Hydraulic systems will bring up cold water and sea floor sediment from the depths and disperse them over the surface waters. Mechanical systems will disperse a sediment plume from the bottom to an indeterminate height, and both systems will cause overturning of sediments in place. In the exploration phase, then, measurements will have to be made of baseline conditions and populations. Disturbances will need to be monitored, and postmining monitoring will have to be maintained for about a year. With the current lack of operations to allow for postoperational monitoring, it may be possible to acquire some useful data from the examination of previous exploration areas. This would require a careful search to locate previous sites of experimental nodule mining.

Operational

For all mining activities at sea, whether exploration or exploitation, there are certain basic operational prerequisites. These include a knowledge of the weather, a platform from which to carry out the mission, a source of power and a knowledge of precise location. Lastly, the ability to maintain position at any desired location is mandatory for the accomplishment of most activities requiring contact with the sea floor.

The state-of-the-art for each of these functions varies, in some cases, being well advanced as in navigation and, in some cases, being in a more primitive state as in environmental forecasting. What sets these functions apart is that their technological development will not be dependent on the viability of mineral resource exploitation, and the costs of innovation will not be a charge against the minerals resources base.

BENEFITS OF MINERAL DEVELOPMENT

The following information is taken from PB 226 005.

The benefits to be gained from the mining of manganese nodules are not limited to the mineral values alone. Table 5.25 enumerates an estimated annual value of benefits from the world offshore mineral development. The numbers are less important than the concept of external benefits that the table presents.

TABLE 5.25: BENEFITS FROM WORLD OFFSHORE MINERAL DEVELOPMENT, ESTIMATED ANNUAL VALUE

BENEFITS	$ \times 10^9$
Direct revenue to government (Bonus, rental, royalty, tax)	3
Raw mineral production	5
Marine services and supply industries (Geophysical survey, drilling, pipelines, tankers, etc.)	5
Ultimate uses of marine minerals (Chemical and manufacturing industries, electric power, transportation, etc.)	100
Application of new technology to other industries	*
Intangible benefits to society (Developing skills and knowledge)	*

*Not estimable.

These externalities will vary between commodities and could have considerable influences on final benefit/cost ratios. In the case of manganese nodules, the application of new technology is immediately identifiable for working at depths of 12,000 feet and beyond.

The ultimate use of sand and gravel, for example, would involve much less of an industrial base than would cobalt, nickel and copper derived from the nodules.

LEGAL CONSIDERATIONS

Existing rules of international law were discussed by Myron H. Nordquist, Office of Legal Adviser, U.S. Department of State at the Symposium on Manganese Nodule Deposits in the Pacific held in Honolulu, Hawaii on October 16-17, 1972 and issued as PB 218 948 in which he posed the threshold question, What area constitutes the "deep seabed?" Limits questions such as this have been the most troublesome in law-of-the-sea negotiations.

Under the Continental Shelf Convention, coastal states have exclusive exploration and exploitation rights over the natural resources of adjacent submarine areas. These rights extend out to the 200-meter isobath or beyond that, to where the depth of the superjacent waters admit of exploitation. The relevant legal issue is: To what extent does the term "adjacency" limit the exploitability criterion? There are sophisticated arguments which can be made to support either a narrow or a wide interpretation of what is "adjacent." It is sufficient to note that by the inclusion of the words "adjacency" and "Continental Shelf" in the Continental Shelf Convention, national jurisdiction was not intended to extend indefinitely seaward. That is, existing law encompasses a deep-seabed area beyond coastal-state exclusive natural-resource jurisdiction.

Assuming that the seabed area under examination is beyond these limits of national jurisdiction which, in the absence of seabed treaty law such as the Continental Shelf Convention, customary rules apply. Evidence of international custom is found in general community practices accepted as law. Unfortunately, there are no direct precedents for clear guidance as to national practice, because manganese nodules have never been commercially mined. However, the deep seabed is subject to general high-seas principles.

The Convention on the High Seas, which is generally declaratory of established principles of international law, confers rights and imposes duties on high-seas users. The foremost rule is that no nation may assume sovereignty over high-seas areas. Specific freedoms such as navigation, fishing, laying of submarine cables and pipelines and ". . . .others which are recognized by the general principles of international law," are provided. The exercise of these freedoms, however, is subject to the condition that reasonable regard be given to the interests of other nations in their exercise of the freedom of the high seas.

What relevance is high-seas doctrine to manganese nodule miners? In brief, it means that until accepted rules emerge governing the miner's activities, he has an international obligation to pay "reasonable regard" to other high-seas users, including navigators and fishermen. For example, irresponsible degradation of the marine environment from his activities would be prohibited. The criterion of reasonableness must be measured against the facts of the particular case. Firmer guidelines would evolve with experience. This process of case-by-case legal development in international law is well known to the common law.

THE GENEVA CONVENTIONS ON THE LAW OF THE SEA

The 1958 Geneva Conventions on the law of the sea form a major element in the current sea law, and are themselves the product of centuries of gradual development. The Conventions were adopted following the first United Nations Conference on the Law of the Sea, which was held that year. The agreements have been ratified by less than half the nations of the world and other countries have not accepted them. Many believe they are in need of revision. Critics of the conventions have observed, for example, that the conventions reflected the interests of the major maritime powers whose voices predominated at the 1958 Conference, and did not adequately take account of the needs of developing countries

The Geneva Conventions are as follows: The Convention on the Territorial Sea and the Contiguous Zone provides that the sovereignty of a State extends to a belt of sea adjacent to its coast, described as the territorial sea. Sovereignty also extends to the air-space above and the seabed and subsoil below but agreement was not reached on the width of the territorial sea. The articles of the Convention say how the territorial sea is to be measured. Ships of all States are assured of the right to pass through territorial sea so long as passage is "innocent" and "not prejudicial to the peace, good order or security of the coastal State". The contiguous zone is an area that may not extend beyond 12 miles from the base-line (generally the coastline) from which the width of the territorial sea is measured.

In this area the coastal State may exercise the control necessary to prevent infringement of customs, fiscal, immigration or sanitary regulations within its territory or territorial sea.

The Convention on the High Seas defines the high seas as all parts of the sea not included in the territorial sea or in the internal waters of a State. It provides for freedom of navigation, fishing, laying pipelines and cables and overflight. The Convention says that all States, whether coastal or not, have the right to sail ships on the high seas and that States not having sea-coasts should have free access to the sea.

Conditions are set for the flying of national flags on ships; there are provisions to ensure safety at sea with regard to signals, labor conditions and seaworthiness; procedures to be followed in the event of collisions at sea are detailed.

States are required to prevent and punish people responsible for the transport of slaves in ships and States "shall cooperate to the fullest extent in the repression of piracy on the high seas". Provision is made for the "hot pursuit" of ships alleged to have violated the laws of coastal States within the State's internal waters or territorial sea. Signatory States are required to draw up regulations to prevent pollution of the seas from ships, pipelines or from exploration or exploitation of the seabed and its subsoil.

The Convention on Fishing and Conservation of the Living Resources of the High Seas provides that all States have the right for their nationals to engage in fishing on the high seas, but States also have a duty to adopt measures for the conservation of the living resources of the high seas.

The Convention explicitly states that conservation of the living resources of the high seas "means the aggregate of the measures rendering possible the optimum sustainable yield from those resources so as to secure a maximum supply of food and other marine products (for human consumption)". It provides that where nationals of two or more States are engaged in fishing the same stocks of fish in areas of the high seas, the States shall negotiate agreements for their nationals to take the necessary measures for the conservation of the living resources affected.

The Geneva Convention on the Continental Shelf says the term is used as referring "(a) to the seabed and subsoil of the submarine areas adjacent to the coast but outside the area of the territorial sea, to a depth of 200 meters or, beyond that limit, to where the depth of the superjacent waters admits of the exploration of the natural resources of the said areas; (b) to the seabed and subsoil of similar submarine areas adjacent to the coasts of islands". The convention states that "the coastal State exercises over the continental shelf sovereign rights for the purpose of exploring it and exploiting its natural resources".

It also states that "the rights of the coastal State over the continental shelf do not affect the legal status of the superjacent waters as high seas, or that of the air space above those waters". The exploration of the shelf and the exploitation of its natural resources "must not result in any unjustifiable interference with navigation, fishing or the conservation of living resources of the sea, nor result in any interference with fundamental oceanographic or other scientific research carried out with the intention of open publication".

The fifth convention is formally known as the Optional Protocol of Signature Concerning the Compulsory Settlement of Disputes. It provides that "disputes arising out of the interpretation or application of any Convention on the Law of the Sea shall lie within the compulsory jurisdiction of the International Court of Justice".

A second United Nations Conference on the Law of the Sea was convened in 1960 to consider further the questions of the breadth of the territorial sea and fishery limits, questions left unsettled by the first Conference. The Conference failed to adopt any substantive proposal on the two questions before it.

The current concern of the United Nations regarding the seabed was inspired by the representative of Malta at the 1967 session of the General Assembly who proposed international action to regulate uses of the seabed and to ensure that its exploitation would be for peaceful purposes only and for the benefit of all mankind.

RECENT INTERNATIONAL DEVELOPMENTS

The following information was taken from COM72 10897.

One way to understand international political developments related to the world ocean and its underlying seabed is to consider the resolutions which were passed by the United Nations General Assembly in the context of the past few years.

In 1967, the UN Seabeds Committee was created by a resolution of the General Assembly. There was an interest in assuring that ocean space would be restricted to peaceful uses. But there was an economic interest too. The less developed countries were determined to take full advantage of the seabed in the hope that its vast wealth would provide the economic equalizer to put them on a par with the industrialized nations.

However, when facts about the extent and character of the wealth of the seabed became more readily available, it became apparent that there was not a great treasure trove of gold, diamonds, or silver lying exposed on the ocean floor, ready to be plucked. It was generally felt that the prospects for black gold (that is, petroleum) were less than slight on the deep ocean floor. Moreover, it was believed that almost all potential petroleum and natural gas reserves existed well within the potential limits of national jurisdiction.

It was further discovered that the only immediate prospects for exploitable wealth on the deep ocean floor beyond the potential limits of national jurisdiction were little black nuggets called manganese nodules, which are rich in copper, cobalt, nickel, and manganese. But it was also learned that the technology to recover or smelt them profitably had not yet been developed. It also became generally accepted that if anyone was on the verge of the technological breakthrough necessary to recover profitably the manganese nodules it was likely to be someone from one of the large highly industrialized nations and not a small less developed country. The less developed nations feared that the developed maritime states would soon produce the needed technology and would then exploit the seabed leaving nothing for them.

Discussion and consensus by member States in 1967 led to the establishment of an Ad Hoc Committee to Study the Peaceful Uses of the Seabed and the Ocean Floor beyond the Limits of National Jurisdiction. The 35-nation Committee informed the Assembly that the item required further study, so in 1968 the Committee on the Peaceful Uses of the Seabed and the Ocean Floor Beyond the Limits of National Jurisdiction was established with a membership of 42 States. The Committee, known as the Seabed Committee, was later enlarged to include the representatives of 91 States by 1971.

Extensive preparatory meetings were held by the Committee and its three subcommittees and working groups in an effort to prepare draft texts on the various issues which would be before the Conference. The Seabed Committee was dissolved by the General Assembly in 1973 prior to the opening organizational session of the Conference in New York in December.

The decision to convene a third Conference on the Law of the Sea was formally made by the General Assembly in 1970. Resolution 2574A of the 24th General

Assembly in December 1969 concerning the seabeds requested the Secretary General to. . . .ascertain the views of member states on the desirability of convening at an early date a conference on the law of the sea to review the regimes of the high seas, the continental shelf, the territorial sea and contiguous zone, fishing and conservation of the living resources of the high seas, particularly in order to arrive at a clear, precise and internationally accepted definition of the area of the seabed. . .beyond national jurisdiction, in light of the regime to be established for the area.

This resolution was carried by a vote of 65 for, 12 against, with 30 abstentions.

The second resolution passed by the General Assembly in December 1969 requested the Seabeds Committee "to expedite its work of preparing a comprehensive and balanced statement of (legal) principles and to submit a draft resolution to the General Assembly at its twenty-fifth session" in 1970. The resolution also encouraged the Seabeds Committee to formulate recommendations regarding the economic and technical conditions and the rules for the exploitation of the resources of the seabed in the context of the regime to be established.

The third seabeds resolution adopted by the General Assembly in 1969, called for the Secretary General to prepare a further study of various types of international machinery, particularly machinery having jurisdiction over peaceful uses of the deep seabed including power to control all activities relating to exploration and exploitation of seabed resources.

The Secretary General had already prepared a comprehensive study on possible forms of machinery, which included the models of a registry system, a licensing system, and an international operating agency. The fourth resolution, the so-called Moratorium Resolution—passed by the General Assembly in 1969, declared that pending the establishment of an international regime, states and persons, physical or juridical, are bound to refrain from all activities of exploitation of the resources of the area of the seabed and ocean floor and the subsoil thereof, beyond the limits of national jurisdiction.

In March 1970, the UN Seabeds Committee resumed work on the development of a set of legal principles in response to the request of the 24th General Assembly. The result was disharmonious with the polarity between the developed maritime powers and the less developed newly emerging nations appearing insurmountable.

Then, on May 23, 1970, the President of the United States issued an historic ocean policy proposal in which it was proposed that perhaps it would be better if the wealth of the seabed could be shared equitably between developed and underdeveloped states alike.

Article I of the Nixon Proposal, after establishing the outer boundary of national jurisdiction at the 200 meter isobath, declares that the resources of the international seabed should be used for peaceful purposes to the benefit of all mankind and especially to the benefit of developing countries. Exploration of the international seabed area would be licensed and controlled by an International Seabed Resource Authority (ISRA), the basic elements of which are included in chapter IV of the Proposal. The area between the 200 meter isobath and the base of the continental slope would be controlled and administered by the coastal state as a trustee for the international community.

Licenses could be obtained for exploitation and exploration, and once exploitation began, work requirements would be established. A portion of the profit from such exploitation would accrue to the benefit of developing countries. Controls on the use of the seabed would prevent unjustifiable harm to the environment. As proposed by the draft treaty, the International Seabed Resource Authority would have an assembly, a council, a secretary general, and a tribunal. ISRA would control and promote the use of the seabed's resources (both mineral and nonmineral) and coordinate with the United Nations in areas of common concern.

Then in the 25th General Assembly resolution 2749, commonly termed the "legal principles" resolution laid down the areas of agreement concerning the status and future of the deep seabed. Briefly it declares:

That there is an area of the seabed beyond the limits of national jurisdiction and that that area and its resources are the common heritage of mankind.

That no nation may appropriate the area to its own national jurisdiction or exercise sovereignty or sovereign rights over it.

That no nation may acquire rights to the area inconsistent with the regime to be established for it and that such activities will be governed by the international regime to be established.

That activities of states in the area shall be in accordance with applicable principles of international law.

That exploration of the area and the exploitation of its resources shall be carried out for the benefit of mankind as a whole, irrespective of the geographical location of states and shall take into consideration the interests and needs of developing countries.

That the area shall be used exclusively for peaceful purposes.

That a regime applying to the area including international machinery shall be established by an international treaty of a universal character.

That the regime shall provide for the orderly and safe development and rational management of the area and its resources and for equitable sharing by states in the benefits derived therefrom.

That states shall promote international cooperation in scientific research in the area.

That states shall take appropriate measures for and provide for the adoption and implementation of international rules, standards and procedures for the prevention of pollution and conservation of the seabed's natural resources.

That in their activities in the area states shall pay due regard to the rights and legitimate interests of the coastal states in the regime of such activities and that coastal states shall be consulted with respect to exploration and exploitation activities with a view to avoiding infringement of their rights and interests.

That nothing in the resolution shall affect the legal status of the waters superjacent to the area or the air space above, or shall affect the rights of coastal states with respect to taking measures to prevent, or eliminate grave and imminent danger to their coastlines arising from pollution or other activities in the area.

That there will be liability for damage caused by activities carried out pursuant to the regime to be established.

That parties to any dispute relating to activities in the area and its resources shall resolve such dispute by measures set forth in Article 33 of the UN Charter; that is by negotiation, inquiry, mediation, conciliation, arbitration, judicial settlement, and other measures—and by whatever measures are agreed upon in the regime to be established.

The vote in support of the resolution was nearly unanimous; the vote was 108 for, zero against, and 14 abstentions. The dispute over recognition of the 1969 moratorium resolution was resolved by compromise. The 1970 resolution on legal principles does not tacitly refer to any prohibition on exploitation, nor does it specifically affirm the high seas freedom to exploit the deep seabed. The United States position is that under international law there is a present right to exploit the deep seabed and indeed prior to establishment of a deep seabed regime.

The resolution makes provisions for inclusion of appropriate international machinery in the regime to be established. Another development reflected in the resolution was the reemergence of a recognition of the special rights of coastal states. Three other resolutions were passed by the 25th General Assembly, one of which, Resolution 2750, calls for the Secretary General of the United Nations, in cooperation with various UN agencies, to do a study for the Seabeds Committee which would:

(a)   Identify the problems arising from the production of certain minerals from the area beyond the limits of national jurisdiction and examine the impact they will have on the economic well-being of the developing countries, in particular on prices of mineral exports on the world market;

(b)   Study these problems in the light of the scale of possible exploitation of the seabed taking into account the world demand for raw materials and the evolution of costs and prices. . .

The next Resolution, 2750B, requested the Secretary General of the United Nations to do a study for the Seabeds Committee on the special problems of landlocked countries relating to the exploration and exploitation of the resources of the deep seabed. The final resolution passed by the United Nations General Assembly was 2750C. It called for a law-of-the-sea conference in 1973 to deal with the

establishment of an equitable international regime—including an international machinery—for the area and the resources of the seabed and the ocean floor and the subsoil thereof beyond the limits of national jurisdiction, a precise definition of the area, and a broad range of related issues including those concerning the regimes of the high seas, the continental shelf, the territorial sea (including the question of its breadth and the question of international straits) and contiguous zone, fishing and conservation of the living resources of the high seas (including the question of the preferential rights of coastal States), the preservation of the marine environment (including, inter alia, the prevention of pollution) and scientific research. . .

It also expanded the Seabeds Committee from 42 to 86 members and charged it with the job of preparing draft articles for a seabed treaty and preparing a comprehensive list of subjects and issues relating to the law of the sea for discussion at the 1973 Law of the Sea Conference. It also provided for review by the General Assembly of the progress of the Seabeds Committee in preparing for the conference at its fall 1971 and 1972 sessions, with the right to postpone the 1973 conference if the progress of the Seabeds Committee were insufficient.

CARACAS SESSION OF THE THIRD UNITED NATIONS CONFERENCE ON LAW OF THE SEA

The following information is taken from Press Release SEA/150 of 30 August 1974, a round-up of the Caracas session.

A start towards a new international convention that would lay down an agreed body of legal rules for the world's oceans was made at a 10-week session in Caracas of the Third United Nations Conference on the Law of the Sea, 20 June to 29 August.

Delegates from 138 nations took part in the session, presenting a wide range of views and seeking to bring together their varying national positions in a set of draft articles for the new convention. After hundreds of formal and informal meetings, three main Committees of the Conference produced preliminary texts of more than 250 draft articles or provisions, many of them in competing alternative versions. To resolve the many remaining differences, the Conference recommended to the United Nations General Assembly, which convened it, that a further session of up to eight weeks be held at Geneva, from 17 March to 3 or 10 May.

In response to the views of a number of representatives who wished to recognize Latin America's contribution to new trends in international maritime law and the arrangements made by Venezuela for the session just ended, the Conference agreed that its final session—for which it did not specify a date—should be held in Caracas for the purpose of signing the documents which will emerge.

In a statement to the Conference summing up its results, its President, H. Shirley Amerasinghe (Sri Lanka), declared: "There has so far been no agreement on any final text on any single subject or issue, despite the lengthy deliberations in the Seabed Committee that formed the prelude to our discussions in the Conference itself. We can, however, derive some legitimate satisfaction from the thought that most of the issues or most of the key issues have been identified and exhaustively discussed and the extent and depth of divergence and disagreement on them have become manifest."

After a series of general statements (28 June - 15 July), in which 115 countries presented their views on the main issues before the Conference, the three main committees, each composed of representatives of all countries participating in the Conference, worked on treaty texts covering the subjects assigned to them, the results of which were the following.

The First Committee, concerned with a legal regime (body of rules) and machinery (world-wide authority) for the area of the seabed beyond the jurisdiction of individual States, produced in informal meetings a revised set of 21 draft articles on a future seabed regime. Under its auspices, negotiations began on what were identified as the main unresolved issues—what entities (such as Government agencies, private firms or the authority itself) should be entitled to explore and exploit the area for seabed minerals, and under what terms and conditions. Still to be tackled are the structure and functions of the seabed authority.

The Second Committee dealt with a wide range of sea-law issues from the terri-

torial sea and the proposed economic zone to the rights of land-locked countries
and the special problems of archipelagos. It set down in a series of working papers some 230 provisions, drafted in treaty language and reflecting delegates' proposals, on the main items before the Committee. The Committee began a second
reading of some of those papers, seeking to narrow down areas of disagreement
and reducing the number of alternative versions. Ahead of it lies the task of
reaching agreement on what many representatives have referred to as a "package
deal" accommodating the varied interests represented at the Conference.

The Third Committee, in informal meetings, worked out texts for draft articles
on preservation of the marine environment and on marine scientific research. Regarding environmental matters, its texts include fully agreed articles on technical
assistance and on the obligation not to transfer pollution from one area to another, as well as five other draft articles with varying numbers of alternatives or
amendments; however, it did not have time to deal with the crucial issue of
standards, jurisdiction and enforcement of antipollution rules.

On research, it agreed to articles on general principles for the conduct and pro-
motion of research and on international cooperation in that area; but on the key
issues of the right to conduct research and the need for coastal State consent to
research conducted off its shores, the Committee was able only to reduce the
number of alternative proposals from five to four.

The drafting work of the Committee was carried on in informal, closed meetings
and did not reach the stage of formal decision. In all three Committees the
drafting and negotiating process was preceded by debates at which delegations
presented their positions. The work of the Conference to date was not reviewed
in plenary meetings of the Conference. However, each Committee produced a
brief statement of its activities (taking the place of a report), and the Chairman
of each summed up the results to date in an oral statement to his Committee.
The President of the Conference made an overall statement at the final meeting.

Following are summaries of the Committees: The First Committee (Seabed Regime and Machinery), at 17 formal and 23 informal meetings between 10 July
and 27 August, concentrated on the future legal regime to govern the seabed
area beyond national jurisdiction. This was the first of the two main aspects before it, the other being the machinery of the proposed international seabed authority.

In a closing assessment of the First Committee's work at Caracas, its Chairman,
Mr. Engo, declared that the "central issue" before the Committee . . . who may exploit the seabed, is more ripe for negotiations now than it ever was. Beginning
its work in the face of divergent opinions and "monumental difficulties", he
said, the Committee had set in motion the processes of negotiations and had
"continued to remove obstacles". Its work in "tidying up" the text of the 21
draft articles on a seabed regime "exposes more than ever the main issues which
must be nogotiated".

The Second Committee (General Aspects of Sea Law) was assigned 15 of the 24
items before the Conference, covering the legal regimes that are to be created for
various ocean spaces from the territorial sea out to the high seas, as well as the
special interests and needs of particular groups of countries such as land-locked
States and archipelagos.

In a concluding statement giving his personal views on the Committee's work, the Chairman, Mr. Aguilar, stated: "The idea of a territorial sea of 12 miles and of an exclusive economic zone beyond the territorial sea up to a total distance of 200 miles is, at least at this time, the keystone of the compromise solution favored by the majority of States participating in the Conference.

"The acceptance of this idea is, of course, dependent on a satisfactory solution of other issues, especially the issue of passage through straits used for international navigation, the outermost limit of the continental shelf and the actual retention of this concept, and last, but not the least, the aspirations of land-locked and other countries which, for one reason or another, consider themselves geographically disadvantaged. There are, in addition, other problems which must be studied and solved in connection with this idea, for example, those relating to archipelagos and to the regime of islands in general.

The Third Committee (Marine Environment, Research and Technology) was able to agree at its informal meetings on some draft articles concerning two of the main topics assigned to its—preservation of the marine environment and marine scientific research.

Regarding preservation of the marine environment, it reached agreement on a text setting out the obligation of States to "promote programs of scientific, educational, technical and other assistance to developing countries for the preservation of the marine environment and the prevention of marine pollution", as well as to "provide assistance, in particular to developing countries, for the minimization of the effects of major incidents which may cause serious pollution in the marine environment".

The text also provides that, to prevent marine pollution or minimize its effects, developing States should be granted preference in the allocation of funds and technical assistance by international organizations and in the utilization of their specialized services. Another draft article agreed to informally provides that, "in taking measures to prevent or control marine pollution, States shall guard against the effect of merely transferring, directly or indirectly, damage or hazard from one area to another or from one type of pollution to another".

There was also agreement on four paragraphs of an article on global and regional cooperation, providing that a State learning of imminent or actual damage from pollution must notify other affected States, and competent international organizations must cooperate in eliminating the effects of pollution and preventing or minimizing the damage from such incidents. States would also be required to cooperate in studies and data exchange and in working out scientific criteria for rules and practices to prevent marine pollution.

Agreement was also reached on one paragraph of a text on "particular obligation", which would require States to "take all necessary measures to prevent, reduce and control pollution of the marine environment from any source using for this purpose the best practicable means at their disposal and in accordance with their capabilities".

The Committee Chairman, Alexander Yankov (Bulgaria), in a personal summation of the Committee's work given to the final plenary meeting, said the unresolved issues related in general to "the scope and extent of coastal State jurisdiction and the rights and duties of other States with regard to marine pollution control and marine scientific investigation". Regarding environmental issues, he said existing agreement in that area "seems to be confined to only a few texts which, by their nature, need to be supplemented or qualified with more specific provisions where consensus remains as yet elusive".

In the area of marine pollution from land-based sources, dumping of wastes at sea and activities on the seabed, "the issues are more clearly set out, and the solutions offered in individual proposals are less sharply divided than would seem to be the case with regard to pollution from vessels". Not yet discussed were such issues as responsibility for damage, immunities and the settlement of disputes.

In view of this seeming lack of progress on the resolution of these problems by the UN, Francis M. Auburn, in a paper presented at the Symposium on Manganese Nodule Deposits in the Pacific held in Honolulu, Hawaii on October 16-17, 1972 and issued as PB218 948 stated that when considering the fast-approaching exploitation stage for manganese nodules, the date of the Law of the Sea Conference becomes an essential matter.

Considering the UN timetable the seabed minerals provisions would probably come into force, at the earliest, in 1980. Well before then, in the current views of active ocean miners, exploitation will be in full swing. With regard to manganese nodules time has almost run out for the UN Seabed Committee. The focus of interest has thus shifted to national practices. This is exemplified by the American Mining Congress' Deep Seabed Hard Minerals Bill introduced by Senator Metcalf.

U.S. SENATE BILL

A talk presented by Leigh Ratiner at the Symposium on Manganese Nodule Deposits in the Pacific held in Honolulu, Hawaii on October 16-17, 1972 and issued as PB 218 948 concerned public policy and the debate on the U.S. Senate bill, the avowed purpose of which is to promote the conservation and orderly development of the hard-mineral resources of the deep seabed beyond the limits of national jurisdiction pending adoption of an international regime by treaty.

Its actual purpose is to establish promptly a legal and operational climate which will give United States mining companies guarantees sufficient to justify major investments in deep-sea mining, and reduce to an inconsequential degree the political risk which results from international negotiations which may alter the legal status of the deep seabed from what it is today—an area subject to the high-seas regime—to another more restrictive international regime yet to be negotiated.

The basic concept is that manganese nodules can be mined under the existing international law of the sea, with an interim regime which would develop customary international law. The Bill covers only the "deep seabed," seaward of the Continental Shelf, but does not define the deep-seabed/Continental Shelf boundary. Under the Bill U.S. nationals could only develop the seabed under license. A licensee would gain exclusive use of the mine site for a 15-year development period.

Exploitation would be permitted so long as commercial recovery continued. An international registry would be created, but only as a clearinghouse. The minimum annual expenditures to prohibit speculation and claim freezing would be $1,350,000 over a 15-year period for each licensed block. An escrow fund, formed of a percentage of United States license fees, would be distributed to developing nations having similar legislation or practice, as Congress would direct. The United States government would reimburse for increased costs or lost assets over 40 years resulting from any future international regime.

The mining industry is reluctant to spend 250 million dollars for each company to develop mining systems, metallurgical processes and prototype and production plants tailor-made for a mineral resource which as a result of negotiations could conceivably become legally inaccessible.

Industry's principal arguments in favor of the senate bill are the following: First, the United States is a mineral-importing country—at least with respect to the minerals found in manganese nodules. While there is not a metal shortage, industry believes, particularly in light of the proclivity of some developing countries for expropriation of investment and control over production and prices, that it is in the U.S. interest to reduce American dependence as much as possible on foreign sources of supply.

Second, industry apparently believes that the Law of the Sea negotiations will take several more years to complete and possibly, it will be an additional five to ten years after that before the treaty has the requisite ratifications to come into force. This delay, will make it necessary to put off investment for so long that capital will move toward land mining rather than ocean mining, and the capital-intensive, technological effort to create this new industry will not take place.

They fear that other countries will engage in the enterprise anyway, with governmental assistance and encouragement. The result will be that by the time a treaty comes into force, others will have a substantial technological, prospecting, and site-identification lead.

On the other hand, there are cogent arguments that the Law of the Sea treaty, when completed in the next several years, can be implemented immediately by national governments, provided they allow for the requirements of the treaty when it comes into force. The United States need not await the actual treaty before implementing it. The United States has not accepted the moratorium resolution of the United Nations General Assembly, and holds the view that until a new regime comes into force, they are entitled to mine the seabed under the high-seas convention. Hence, legislation closely geared to the treaty could be enacted once signed, which would give industry the necessary assurances so that they could make their investments. Moreover, the short delay inherent in this approach would probably not result in the substantial risk of foreign countries closing the technological gap.

Third, industry has closely watched the UN Seabeds Committee and has noted that a few countries (whose economies are closely related to the mining of certain of the metals found in manganese nodules) are intensively pushing for an international regime which can control both production and prices of minerals through an organization that would be dominated by developing countries and legally empowered to have the exclusive authority to mine and market seabed minerals.

In the interest of "developing country solidarity"—an important political reality in UN politics—many other countries are giving tacit support to this international regime approach. Industry concludes then that, even if negotiations could be speeded up to meet their timing needs, the result will be unacceptable to them such industry fears that it may be necessary for the United States to accept a fundamentally harmful treaty on the deep seabeds in order to obtain other objectives in the treaty, such as the freedom of transit of U.S. warships through international straits.

Fourth, industry holds the view that if it has a right to mine the seabeds under present international law, it is patently unfair to ask it to mine while the United States is negotiating a treaty which will almost certainly do away with the regime of the high seas for deep-sea exploration and exploitation. As stated above, investing 250 million dollars knowing the law may soon require surrender of mine sites which justified the investment, would be regarded as a rather poor investment decision.

REGULATION OF HARD MINERAL MINING ON THE CONTINENTAL SHELF

The following information was taken from COM72 10897.

The world's continental shelves, covered by far shallower waters than the deep seabed, and much closer to land, also contain minerals. The most famous mineral of the continental shelf is, of course, petroleum. The continental shelves are also storehouses for deposits of such important hard minerals as diamonds, gold, tin, iron, chromite, ilmenite, magnetite, and sand and gravel. It cannot yet be said that there is an "offshore hard mineral industry" on the continental shelves in the same way that one can point to the "offshore oil industry." Some sand and gravel is being mined off the coasts of the United States and a few other minerals are being taken from the shelves in other parts of the world.

While these activities remain in the venture stage, some consideration should be given to the technological aspects of mining the continental shelf since: The offshore sand and gravel gross economic value for New York in 1967 was $20 million; in California sand and gravel value was $0.5 million; diamonds have been mined along the Atlantic coast of SW Africa but not economically; iron has been dredged off the southern tip of Kyushu Island; gold has been found in offshore Alaska but not in concentrations high enough to be economic; and tin in offshore Thailand is being explored.

Total gross value of world offshore hard-mineral mining operations in 1967, excluding sulfur which like oil is mined in a liquid form, and excluding iron ore and coal operations that are conducted from onshore locations with traditional mining methods, was only $180 million; of this amount $100 million was accounted for by sand and gravel. There are however, more and more indications that new technology will allow these ventures to pay off. It has been appropriate for several years to expend time and energy propounding and discussing management proposals for deep sea mining; it is surely time (from a technology viewpoint) to talk about management schemes for mining the continental shelf.

There are significant differences between the two areas. Most importantly—at least since the 1958 Geneva Conventions on the Law of the Sea—the continental shelves have been within the limits of national jurisdictions and thus subject to existing governing authorities. Much of the controversy over the deep seabed stems directly from the fact that it lies outside the limits of any nation's recognized jurisdiction. In the United States, the coast states own the resources of the continental shelf out to three nautical miles from shore (for historical and technical reasons Texas and Gulf-side Florida have seaward boundaries extending nine nautical miles from shore). The federal government owns and administers the lands lying beyond these limits to the edge of the legal continental shelf.

There are, counting the states on the Great Lakes, thirty states whose borders are formed to some extent by the sea. Each state has the authority under the Submerged Lands Act to regulate the profit from the extraction of hard minerals (among other resources) from its portion of the continental shelf. Of those thirty states, twenty can claim some sort of legislative or administrative reference to offshore mining of hard minerals. (These laws can be found in the Appendix of COM72 10897.) But of those twenty references, ten exist only as small parts of larger schemes mainly concerned with upland mining, two are apparent afterthoughts of offshore petroleum legislation and six apply only to sand and gravel.

That leaves only two states, Alaska and California, which have so far taken the trouble to construct "comprehensive" management patterns especially for hard-mineral mining in offshore areas.

OIL SPILL PREVENTION
AND REMOVAL HANDBOOK 1974

by Marshall Sittig

Pollution Technology Review No. 11

Energy Technology Review No. 2

Ocean Technology Review No. 1

This handbook describes and expounds the most feasible and sophisticated technologies available at present for dealing with a menace of worldwide proportions, i.e. marine pollution caused by accidental contamination with crude or (less often) refined petroleum oils. To make matters worse the crude oil has a specific gravity very close to that of seawater.

Oil slicks, tar balls, and dispersed hydrocarbons are found on the surface as well as on the bottom of the ocean. Very little has been accomplished so far about relieving the latter calamity. Once the spilled oil has found something solid to cling to—it may be the sand of a beach, underwater rocks, wooden structures, the feathers of a gull, or the hair and skin of a bather—it does not readily let go.

Aside from most rigorous safety regulations and preventive measures now in force for tankers, offshore drilling, and seabed mining (still in its infancy), the worldwide effort towards combating oil spills is best illustrated by the myriad, albeit rudimentary, devices being designed by companies engaged in the development of containment booms and other mechanical containment apparatus for dealing with oil slicks on the surface of the sea.

Research continues to obtain improved performance, but, because of natural turbulences and gulfstream-type currents, the effectiveness of floating booms and other recovery systems is severely limited.

Once a spill has been contained and the sea is calm, cleanup measures can be instituted. The two major techniques used today are mechanical gathering and recovery with adsorbents. Mechanical recovery or skimming may be used without restriction in carrying out the contingency plan, while adsorbents may be used only if these materials do not by themselves or in combination with the oil increase the pollution hazard. Straw, manufactured fibers, and natural clays may be mixed with the oil slick, and collected. Straw is considered the most

cost-effective agent, because it floats readily even when wet, holds five times its weight in oil and costs little.

Other methods include use of dispersing agents, sinking agents, biological agents and burning. Approval by the appropriate governments is required for each of these measures and is practically never granted for U.S. Coastal Waters. It has been suggested that relatively small beaches can be cleaned by feeding the contaminated sand through the municipal incinerator.

Even a relatively small oil spill may have effects that are catastrophic and long-lasting, but prompt application of the methods described in this handbook will at least help to prevent successive biological and degradative changes of the marine biota and the ecological entity.

In this handbook are condensed vital data from numerous government and other sources that are unusually difficult to find and to pull together. Important oil containment and removal processes are interpreted and explained by examples from recent U.S. patents.

A partial and condensed table of contents follows here. Numbers in parentheses indicate the number of processes or designs per topic. Chapter headings are given, followed by examples of important subtitles.

EFFECTS OF OIL SPILLS ON THE ENVIRONMENT

Effects on Man
Effects on Agriculture
Effects on Industry
Effects on Wildlife
  Waterfowl
  Shellfish
  Fish
  Marine Mammals
  The Marine Food Chain
Effects on Marshes

PREVENTION AND CONTROL OF SPILLS FROM OFFSHORE OIL PRODUCTION

Mud Cleaning
Leakage Control
  at Well Head (23 processes)
Wastewater Purification (3)

PREVENTION AND CONTROL OF SPILLS FROM OIL TRANSPORT

Oil Transfer Operations
  Pre-Transfer Conference
  Transfer Procedures
  Vessel-to-Shore Transfer
  Shore-to-Vessel Transfer
  Miscellaneous Transfer Precautions
  Terminal Facilities
  MacLean Process
  Wittgenstein Process
Tank Cleaning (3)
Ballast Disposal (3)
Oil Removal from Tanker Accidents (4)
Drydock Operations

PREVENTION AND CONTROL OF SPILLS FROM OIL REFINING

Treatment of Oily Wastewaters (3)

PREVENTION AND CONTROL OF SPILLS OF INDUSTRIAL OILS

Treatment of Oily Wastewaters (2)

OIL SOURCE DETECTION, IDENTIFICATION AND MONITORING

Detection and Monitoring (2)
Spill Quantity Estimation
Sampling
Tagging
Identification

ORGANIZATION FOR OIL SPILL CONTROL

The On-Scene Commander
Industry Spill Cleanup Cooperatives
Relations with Government Organizations
Financing
Use of Independent Contractors

OIL CONTAINMENT

Air Barriers (7 designs)
Chemical Barriers (4 processes)
Floating Booms (103 designs)
Combined Containment and
  Removal Devices (18 designs)

OIL REMOVAL FROM OPEN WATER SURFACES

Weir Devices (16)
Floating Suction Devices (14)
Sorbent Surface Devices (32)
Vortex Devices (6)
Screw Device
Jet Devices (2)
Other Devices (4)
Skimming Vehicles (49 designs)

OIL REMOVAL FROM STREAMS, LAKES, HARBORS

Proprietary Processes (6)

OIL REMOVAL FROM UNDERGROUND WATERS

Proprietary Processes (2)

OIL/WATER SEPARATOR DEVICES

Proprietary Devices and Processes (18)

PHYSICAL AND CHEMICAL TREATMENTS OF OIL SPILLS

Dispersion Treatments (3)
Sorption Processes (33)
Sinking Processes (4)
Gelling/Coagulation Processes (8)
Magnetic Separation (3)
Combustion Processes (14)

OIL REMOVAL FROM BEACHES

Spill Control Measures
Chemical Control Agents
Combustion Methods
Culpepper Process
Froth Flotation Treatment

DISPOSAL OF RECOVERED SPILL MATERIAL

ECONOMICS OF OIL SPILL PREVENTION AND CONTROL

FUTURE TRENDS

ISBN 0-8155-0543-4

466 pages

GEOTHERMAL ENERGY 1975

by Edward R. Berman

Energy Technology Review No. 4

This book describes in detail the nature of the geothermal resources, their extent, and the currently available technology by which these natural sources of energy can be exploited and utilized to the greatest advantage. There is little pollution resulting from the use of geothermal energy, and with a little care all disturbance of natural ecological systems can be avoided.

Earth heat can be used most practically where hot volcanic rocks are comparatively near the surface, and emerging circulating ground waters act as heat collectors, either by producing steam or by serving as heat transfer media. Suitable sites have been discovered in the Continental U.S. and Hawaii, the U.S.S.R., Japan, New Zealand, Iceland, and Italy. In view of the rise in the price of oil and the intensive search for new sources of energy, geothermal power appears to be in for a period of rapid development.

This Energy Technology Review is based on international studies conducted by industrial and engineering firms or university research teams under the auspices of various governments and governmental agencies.

A partial and condensed table of contents follows here.

ISBN 0-8155-0563-9

336 pages

INDUSTRIAL WATER PURIFICATION 1974

by Louis F. Martin

Pollution Technology Review No. 16

The Federal Water Pollution Control Act (now made into law) will have far-reaching effects on all industry. The pretreatment standards, about to be published as a result of this act, will exert increasing constraints on industrial process design and expansion plans in the years ahead.

The *Guidelines for the Pretreatment of Discharges of Publicly-Owned Works* prohibit the introduction of industrial pollutants which would pass through a municipal facility "inadequately treated," therefore, all existing industrial facilities and new plant designs must include specific downstream water purification processes, before the effluent is sewered or discharged into public systems.

This book describes over 160 recently devised processes for the treatment of contaminated industrial waters.

A partial and condensed table of contents follows here. Numbers in parentheses indicate the number of processes per topic, chapter headings are given, followed by examples of important subtitles.

While conventional separation processes for solid-liquid and water-oil effluents are applicable to many industrial streams, the process literature clearly highlights the areas of greatest immediate concern as shown by this table of contents:

ISBN 0-8155-0554-X

300 pages

ENERGY
FROM SOLID WASTE 1974

by Frederick R. Jackson

Pollution Technology Review No. 8

Energy Technology Review No. 1

The solid waste disposal problem is reaching alarming proportions everywhere. The United States alone produces close to 300 million tons of solid waste per year, which is equivalent to about one ton per person.

At the present time the prevalent methods of disposal are dumping and sanitary landfill. Municipal incineration disposes of a small portion only, with attendant high capital and operating costs.

Many methods have been proposed for coping with the problem, such as source separation, source reduction, or material recovery. However, with the energy crisis descending upon us, producing energy from waste is becoming more and more attractive. An estimate currently making the rounds in financial circles is that when the price of crude oil reaches $7.00 a barrel, alternate sources of energy become practicable.

This book is based primarily upon information from studies conducted under the auspices of the EPA. Its foremost topic is burning of solid wastes to create steam directly. The air pollution problem created by burning can be solved quite easily with known technology. Solid wastes are low in sulfur, consequently there is no SO_2 removal problem.

Another technique that may assume more importance in the future is the controlled pyrolysis of wastes, yielding so-called pyrolysis gas or oil. Chapter seven is devoted to this. The final chapter discusses European practice, which is historically far more extensive than that of the U.S. A condensed table of contents follows here:

ISBN 0-8155-0528-0

163 pages

DRILLING MUD
AND FLUID ADDITIVES 1973

by John McDermott

Chemical Technology Review No. 20

With the completion of the Lucas gusher at Spindletop, near Beaumont, Texas, in 1901, the rotary method of drilling oil wells became well established, and a drilling mud consisting of water and finely ground cuttings from the hole was employed. To this was later added a quantity of gumbo, a surface clay found nearby, which proved highly beneficial.

Through the 1920's iron oxide and barium sulfate (barite) were used to increase the density of the mud, thus preventing entry of the formation fluids into the bore hole. The use of bentonite clay to suspend the barite forms the basis for today's large commercial drilling mud industry.

Today's high performance drilling muds and fluids make it possible to drill thousands of feet of open hole at substantial savings in casing costs.

This book presents over 160 processes and several hundred examples of typical fluids as described in recent patents. In view of the multifunctional nature of many drilling fluid additives, the book focuses on the major functional contibution of each additive. The use of surfactants as dispersants, suspending agents, and emulsifiers is treated separately, followed by specific chapters relating to formulations developed for fluid loss, loss of circulation lubricity, and the many related problems encountered in this industry.

A partial and condensed table of contents follows. Numbers in parentheses indicate the number of processes per topic. Chapter headings are given, followed by examples of important subtitles.

1. SURFACE ACTIVE AGENTS (34)
Water-Based Fluids
Polyoxyethylene Sorbitan
Tall Oil Esters
Rosin Amine-Alkylene Oxides
Oil-Based Fluids
Salts of Petroleum Sulfonates
Sulfonated Sperm Oils
Surfactants and Diesel Oil
Air Foams
Diphenyl Dimethyl Polysiloxanes
Sulfated Polyoxyethylated Alcohols

2. FLUID LOSS AND VISCOSITY CONTROL (62)
Water-Based Fluids
Cellulosics, Guar Gum, Polyacrylamide Hydrolyte, and Vinyl-Maleic Copolymers
Resin Emulsions
Modified Starches
Lignin Derivatives
Metal Sulfates from Sulfite Liquor
Sulfonated Lignins from Spent Liquors
Copolymers
Hydrolyzed Polyacrylamides
Other Additives
Castor Pomace
Tall Oil Pitch + Polyoxyethylene
Fragmented Asphalt Products
Clay Treatments
Prehydrated Bentonite
Salt Water Muds

3. OIL-BASED FLUIDS (11)
Stearic Acid + Oyster Shell Flour
Inert Absorbents
Humic Acid + Tall Oil + Polyamines
Fatty Alkyl NH_4-Lignosulfonates
Lignin + Amines

4. LOST CIRCULATION AND SEALING MATERIALS (17)
Flexible Flakes and Fibers
Phenol-Treated Wood and Glass Fibers
Diatomaceous Earth + Asbestos
Sand-Combining Liquid Injections
Oil-Dispersible Colloids

5. LUBRICITY AGENTS (18)
Antisticking Additives
Molybdenum Complexes
Sulfonated Tall Oil Pitch
Hard Particles of Uniform Size
Lubricants for Diamond Coring Bits
Slimhole Drilling
Sulfurized Phenolics

6. DEFLOCCULANTS, DEFOAMERS, CORROSION INHIBITORS AND OTHER ADDITIVES (23)
Stabilized Salty Well Fluids
Water-Soluble Phosphate Esters
Addition of Scavengers
Microbe Inhibitors
Reactions for In Situ Thickening
Hydrophobic Liquids
Processing of Low Silt Muds
Thermal Insulating Fluids
High Density Barite

ISBN 0-8155-0510-8

305 pages

SANITARY LANDFILL TECHNOLOGY
1974

by Samuel Weiss

Pollution Technology Review No. 10

This Pollution Technology Control Review surveys all the latest technical information on this subject. It is based primarily on studies conducted by industrial or engineering firms, under the auspices of the Environmental Protection Agency.

Sanitary landfill practice is an engineering method of disposing of solid wastes on land, whether they are of municipal or industrial origin. There is no resemblance to the old-fashioned garbage and rubbish dump. There are no fires, no obnoxious fumes or smoke, no flies and no rodents or other scavengers.

As stated in the introduction to this book, a sanitary landfill is not only an acceptable and economic method of solid waste disposal, it provides also an excellent way to improve the commercial value of otherwise unsuitable or marginal land areas within a few years.

A partial and condensed table of contents follows here.

ISBN 0-8155-0542-6

300 pages

INCINERATION OF SOLID WASTES 1974

by Fred N. Rubel

Pollution Technology Review No. 13

This book gives advanced technical and economic information to promote efficient incineration of municipal refuse, and will help to erase the bad neighbor image of incinerator installations within the community. Special incinerator applications, largely outside the scope of municipal waste disposal, have not been neglected.

Incineration offers the most significant volume reduction of solid organic wastes when compared with all other disposal methods.

The most undesirable side effect is, or rather was, air pollution, and here the primary concern is with emissions of particulates, rather than with gases and odors. Many sophisticated devices for nearly complete retention of these solid particles have been developed in recent years, and special emphasis has been placed upon them in the descriptions of apparatus.

Most of the data and other information presented in this book are based upon detailed reports on government contracts fulfilled by industrial companies under the auspices of the Environmental Protection Agency.

A partial and condensed table of contents follows here. Chapter headings are given, followed by examples of important subtitles.

ISBN 0-8155-0551-5

246 pages